PSALMS OF HORROR

First edition. September 27, 2025.

ISBN: 979-8232281908

Written by Traian Tus.

Table of Contents

PSALMS OF HORROR

And how the world could end tomorrow...

PSALMS OF HORROR

And how the world could end tomorrow...

Traian Rareș Tuș

To my brother,

the most curious person I know

A thought about this book

When an adolescent of genius – the way Traian Rareş Tuş seems to be, originated from Baia Mare, Romania – writes a book about the end of the world, it is, without a doubt, a sign: we are sent a warning, we are asked to pay attention. Whether it is a good or a bad omen is left to be concluded upon close analysis. For the time being, I construe for the lector that must be incited to reading, that it is about an exceptional product of the mind of a youth matured before time and preoccupied by the great threats to this planet, to humankind. Traian Rareş will have the big chance to affirm himself in the most erudite scientific environments in the world and will bring along many remarkable messages. One of them refers to the waking of humanity from "the sleep of reason" and I want to believe that his voice – united with other lucid, responsible voices – will be able to change our dying world for the better. Warm congratulations Traian Rareş Tuş and farewell in the tumultuous life that awaits you!

Academician Ioan-Aurel Pop,

President of the Romanian Academy

PREFACE

"We don't know what will happen tomorrow, but one thing is guaranteed – God's

overarching care for His children. We can be sure enough of that.

In a world where nothing is sure, He is sure."

David Jeremiah

There is truly no telling what tomorrow has in plan for us – we can all agree on this one thing, no matter our religion and beliefs. We will laugh, we will cry, we will love, we will suffer and we will experience a great many other emotions and states, but there is one to top them all off: we will live. And we must live our lives to their fullest, for every moment is a blessing, and this is our way of being thankful for that.

The inability of knowing the future is such a blessing to humans. It constitutes the beauty of life. Were one able to accurately predict every event the future will bring, that person would never be capable of attaining happiness.

But this gift often seems like a curse to us. We didn't accept a simple life. We wanted more than that. It is within the human nature to be curious, to wish for more. Whether we are talking about physical objects, information or any other kind of earthly desire, this most certainly implies that it is also within the human nature to complicate things. We wish, above all, for a happy and easy life, but most of the times we tend to, perhaps unknowingly, work against it.

"Funny how we get attached to the struggle"

Celeste

And it is because of this nature that we urge for knowing more about what's to come. We want to predict the future. It is a very destructive trait on our behalf, but we don't seem to mind. It's as if we weren't happy with today's conditions and expect tomorrow's to be better. This is quite common sense: today is fine, but tomorrow should be great. On a closer analysis, however, we figure out that this is not the case: yesterday ended up being way better that today, even though it didn't feel like it. We ended up not enjoying yesterday enough, because we expected too much of today. Here comes the destructive part – we realized all of that when it was too late, when there was no way to go back and enjoy yesterday, as you already woke up ready for the next day...

It is, indeed, true that sometimes our expectations are met, but in the end, was it worth it? To be so very much looking forward to something that you completely disregarded something else?

And frankly, our suppositions are correct in an astonishingly little percent of the cases. The majority consists of the actual reality defying our presumptions. The events concluded in something worse, or even much worse that what we had previously anticipated.

What is this book about?

This doesn't promise too joyful an atmosphere for the book, and I am aware of that. Truth be told, it doesn't need to. But if you're reading it right now, then you weren't expecting something jolly either.

The subject I intended to tackle is a controversial one, to say the least. Why is that? Well, I believe that, if you're anything like me, this subject must've occurred to you at least once in your life. And you've implicitly looked up for some piece of information to give you assurance or consolation. If not, then you probably will at some point upon finishing this book.

I don't mean to frighten you with the topic I will put forward, but I do wish for you to give it a second thought, to interpret it in your original way, as this is the only means by which you will be faithfully convinced.

So, without spoiling your reading experience too much, I will openly state that this is a book about death, about destruction, about the end of days. In spite of the title, it is not a horror story... it is the harsh, horror reality.

There is a slight twist to all that, so that you will not be having only nightmares from now on. Besides the dark truth of the world we live in, there is also a glimmer of hope, on the other side of the tunnel leading to the abyss. We must first and foremost see that glimmer and persist with our efforts to not fall into the abyss, however likely that would seem.

For this reason, I wanted to begin by studying the end of days as predicted by the Christian Holy Book, the Apocalypse of John from the Book of Revelation. With that foundation, I will elaborate on many possible scientific doomsday scenarios depending of the likelihood of them actually taking place.

Why choose to do this?

I know it sounds a bit out of place up until this point, but hopefully everything will clear up as we advance.

I cannot call myself a very dedicated Christian. As a matter of fact, in my eyes religion itself has brought about more division that unity, even though it wasn't meant to. What I think matters is that I have some kind of belief. I do believe in a supreme being or entity that may or may not have influence over us, but I believe It is there. I am thankful to It and I constantly ask for help or atonement.

This may not seem like enough to some. Alternatively, it might seem like too much to others. I am not rightful to judge anyone's opinion, but in the deep of my heart I know that It helps me.

Based on that, you might find it offsetting that I start my discussion from the Holy Book of a specific religion, my religion. That is justified. But because there is a seed of Christianity inside of me, no matter how big, I figured it would be interesting to relate to it in the first place.

To me, the beauty of the Bible is determined by the numerous interpretations it allows – everyone perceives it differently, and one simply can't say that an interpretation is right and another is flawed. There are at least as many interpretations as there are readers out there. Therefore, I am not trying to infuse you with one "correct" judgement. I am only going to present to you, the reader, one possible interpretation of the biblical apocalypse which seems to suit the science behind it.

Science and belief? How come?

There is a wide variety of books and works by certain religious people. Some of them are inspiring... some of them not so much. But one thing most of them have in common is the partial or total disregard for science. The point of them is to relate on faith and belief and one needn't pay attention if that isn't his or her cup of tea.

On the other end of the spectrum, there are numerous scientific books and papers which focus so much on science (again, this is the point of them), that they straight up deny the existence of divinity.

Many people who go on reading these publications eventually become disappointed – either they are let down by a radical religious interpretation of the world or by the closure that nothing is divine. Some even become convinced that only one of the two ways of thinking is appropriate. I personally find that extremely frustrating.

In my years of activity, though not too many, I have come to the conclusion that science, any branch of it really, is an amazing field. It is a field of possibilities, of information, of that harmful yet enticing curiosity we were talking about. To me, there is no doubt that if you have any questions regarding the aspects of reality, science will most definitely answer them.

From biology, the study of life, to chemistry, the study of matter, and all the way to physics, the study of the universe's constituents, big or small, science is the basis of the entirety of world's wonders. To sum it up, if there's anything certain, science is surely in play, and there is no way for me to treat it with indifference.

The motivation

With the everlasting conflict between science and belief in mind, I figured it was about time someone calmed the waters.

To be clear, I'm not implying that there will not be any doubt left after finishing this book, but I have tried to create somewhat of a common ground between the two throughout this work. This should be an accessible material to all readers, no matter their belief or inclination towards science. To that end, I chose to provoke you to a rather macabre but intriguing discussion about how the world could end, how our species could be obliterated or irreversibly downgraded by a cataclysmic event. Because if there is one thing that binds us all humans is the unpredictability of the future, be it good or, in this case, really bad.

Being constantly terrified by such scenarios myself, but also optimistic of nature, I sincerely hope that you find some soothing in this book. Furthermore, I would be glad if you were to become acquainted with some more obscure corners of science, and especially if you became inspired so that you'll look more into them yourself.

"I am neither clever nor especially gifted.

I am only very, very curious"

Albert Einstein

ACKNOWLEDGEMENTS

The material of this book ranges over multiple fields, theories and interpretations. In consequence, it is not rightful of me to express myself to the readers on such a controversial issue after only six years of experience in the area of science. I believe that when it comes to such issues, it is paramount to put yourself in other people's clothes and see how they look upon the subject. I have thus proposed this topic to some of the most insightful people I know, and I would like to thank them all for baring with my series of interrogations, consultations and stimulating conversations. Those include my family, friends and professors, as well as some of the other mentors I've had contact with over the years – I couldn't have done this without you.

Those being said, a friendly disclaimer might be appropriate before the start of your reading: this is neither a spiritual work, nor a full out scientific one. It is rather a bit of both, with a grain of exegesis from my side. It is an attempt to expose the connection between science and belief by analyzing one of the most asked questions throughout all of history: when will the world end? The book is meant to be a recreational one, but at the same time one that encourages you to reflect more deeply on this matter.

INTRODUCTION

"How do you live in an age of bewilderment, when the old stories have collapsed and no new story has yet emerged to replace them?"

Yuval Noah Harari

I was calmly sitting at the lunch table with my family after getting back from another boring day at school. I was thinking to myself: "Oh, baked beans again?! Can't I just go to sleep and make my homework in the evening to just get over it? I mean, I'm fed up with all this school stuff... why do we even have to go to school? If only there was no school left... it would all be much better!". But man, I couldn't have been more wrong...

It all started when my mother picked up the phone. On the other end there was my father she was speaking to, but something didn't feel right. He is usually as cool as a cucumber, but this time I could hear him yelling at my mother. And it wasn't because he was angry, far from it. He was scared. Terrified even. Yet my mother didn't seem too touched by his words as she continued cooking something for later in the day. Then she became confused. Then angry. Nothing special, but then she screamed in horror when the house started shaking. And it wasn't a gentle shake so that the glass of water trips overs, not even so that the walls crack. It was a violent one, with every appliance permanently damaged, electrical circuits destroyed and the entire house collapsing upon itself. We rushed under the table as terror seized us. I remember how I was screaming and crying when my father rushed in.

I was so happy to see him, there to save us, but naturally I couldn't hold my tears. He was even more frightened then we were. Still, he grabbed us and headed straight towards the car – a company car, a limousine to be precise. Then he started driving, faster than ever, racing through the neighborhood as every other house crumbled. It all got much worse when giant cracks in the streets began to appear. Those huge bumps made me sick... Nonetheless they didn't even come near to the immense chunks of ground that were jumping up and down. The entire city was shattering, standing no chance in front of what seemed to be the biggest earthquake one could possibly imagine.

Gargantuan holes kept on emerging, but we finally made it to the company's runway. He knew a bit of gliding so he tried his luck with the first plane we could find. What else was there to do anyway... He revved up the engine as much as he could before the chasm reached us. In spite of his greatest efforts, we weren't able to achieve too high of a speed until we found ourselves dropping into the ravine...

Yet somehow... we made it! We managed to glide along the steep walls of what used to be Earth's crust and avoid the incoming debris. We glanced back at our former home. It wasn't a city anymore... it was the facsimile of the Great Canyon...

Fortunately, we managed to escape unharmed or, better put, alive. Others were not so lucky. It was only then that we tuned up the radio to find out that this did not affect merely our city, but the entire planet. Coastline cities were now submerged underwater, while others were completely erased of the surface of Earth due to volcanic eruptions.

This was only the beginning. Minutes later, we woke up to an explosion so great that it made a nuclear bomb look feeble. At first, the light was so bright we could barely recover from our temporary blindness. But we wished we stayed that way... in darkness... so we wouldn't see what came next. Fireballs were roaming the skies while ash and cinder were closing in like a giant curtain. As we were frantically moving our head from left to right, we were taken aback by the bolide that was targeting our plane. Another bright light. Then darkness... complete... utter... darkness. No sounds could be heard either, apart from my desperate scream. Was that it? The end of it? Is there nothing that can be done? I am sorry I made that wish! I want to go back to school! I want everything to be normal again!

A bad dream

Well... the story I told you wasn't quite accurate. It wasn't exactly real. But it surely felt real enough to me. To my 7-year-old self, at least, who ended up opening his eyes with his mother wiping the tears of his face and his father patting his back. "It was only a bad dream" they told me. Yet this bad dream has been the most vivid and horrifying one in all my life. To this day I am not certain how that nightmare felt so real, however I do know what caused it.

It was a movie I watched that evening that triggered an entire chain of reactions in my puny mind. Later in my years I have come to appreciate that movie, especially for its mind-blowing special effects. And it was for those effects, which felt so realistic back then, that Roland Emmerich's creation "2012" had forever impacted my childhood.

I didn't recall back then, but I had seen a few of his films before the appearance of "2012" – all depicting doomsday scenarios, some more realistic than others, yet none as realistic as this one. The fright was further amplified by the presumed end of days which was meant to come sometime that year, according to the Mayan calendar.

But what could a young child know? My parents weren't fooled by such claims that kept on coming in every newspaper and TV channel. I wouldn't listen to the news at the time, I don't even do it nowadays. Though at every meal of the day we would either watch TV or listen to the radio, I wasn't able to pay much attention, so that was not a problem.

What I found tremendously disappointing were the cartoons. I remember some of my favorite cartoons being filled to the brim with these kinds of apocalyptic prophecies. Not too explicitly, but assuredly, they made us expect what was to come. And all those started years before 2012. The age of 7 was when I started to realize what things meant – the age when I saw the movie, one year before the end of the Mayan calendar – the age when I stuck my head into the pillow for 30 minutes straight on New Year's Eve so that I wouldn't witness the destruction 2012 was supposed to bring upon us.

That was how it all began...

Another Mayan cycle?!

After this unfortunate experience when I wouldn't get my head out of the pillow (which was the apogee of my heart rate), I had that feeling of relief. The feeling upon winning an argument. The feeling of getting in bed after a tiring day. The feeling of hugging someone after being separated. It was all that and much, much more. I felt like a gladiator who has just won the greatest confrontation of his life, despite me doing nothing as to favorize my winning chances – what was I to do really? But that sentiment of glory was etched into my memory, never to be erased. And it remained conceivably one of the biggest emotions of joy I've ever experienced.

Time went by quite normally afterwards. I was just feeling a bit more accomplished for the next few months. I narrated the events to my classmates to whom I've previously complained about the movie and about the news regarding a worldwide cataclysm and I was happy that they shared my view. But before long, this short-lived happiness was obliterated as another mass destruction event broke the news.

This time it was a giant asteroid that "was going to impact Earth". I've read a couple of books and articles about those and learned about the damage they can cause, even if they come in smaller dimensions. This one, however, was 5 kilometers in diameter, or so they said. Half of the one that wiped out the dinosaurs, but still big enough to terminate us for good.

Being something novel to me, something cosmical, I didn't have too much of an idea of how we can avoid or destroy such a space wanderer, no matter how many things I read. And as expected, the news were only to worsen the already messy situation. Everyone who bit that was completely out of hope. Was there truly no way to survive an incident like this one?

In fact, there was. What was it? The incident being not even near to taking place. As I emerged myself in more and more articles I finally got to the well-documented ones. The scientific articles, written by actual scientists and not some random person in the streets who claims to know better.

The chance of that asteroid hitting Earth was minuscule, insignificant. It was less than 0.01 %. It was a rogue one who happened to enter the inner Solar System after its orbit was somehow distorted. You were more likely of dying of a heart attack while hearing the bad news than of that asteroid impacting our planet. At that point I was convinced it was just 2012 over and over again.

You probably never heard of it again, once those articles made it to the spotlight. What was the most probable cause of another apocalyptic piece of information going haywire? Another Mayan cycle.

Most people started to ignore those after a few crazy attempts of predicting the world's end. Yet there are still incredibly many who believe such things whenever they see them for the first time, without even bothering to question themselves – is it really so or am I fooled by someone aching for attention?

Nowadays, I am somewhat glad for this asteroid scandal bursting out as it was one of the reasons, if not "the reason" I started the pursuit of science, mainly astrophysics. But if you just sit and analyze how often those asteroid scandals occur you'd be shocked. I have been through the end of the world at least half a dozen times that I know of. And not just asteroids. There were also giant earthquakes, super-volcanos you name it. And those are only the widely broadcasted ones, the ones that made it to the entire world's eyes. These doomsday events "happen", on average, once every three years. If you were to count in all the scenarios that pop up on the internet there wouldn't be week without you begging for a few more days.

The issue with the apocalypse

It is a known fact that people do get manipulated or at least influenced on a daily basis – sometimes knowingly, but most of the times without them even noticing. And it has been like this since the dawn of history. It's not about the elite manipulating the masses, it's about the masses not putting forward any resistance to conspiracy theories or simply failing to examine what they've heard.

And then there is one other factor which brings calamity upon our entire species. It is the fact that we people tend to let ourselves much easier influenced by the negative aspects of society than we do by the positive ones. It's as if we are blinded to what's good, to what makes us happy, but the moment we hear word of something bad, our pupils dilate in desire to know more about it. The human nature is a complex mechanism, too complex to be understood at its fullest.

It only goes downhill from here. The worse the news, the more people will pay attention to them and the more adepts will such a theory gain, no matter whether it's proven or not. And what is the worst news one could receive? Well, the answer is quite straightforward – death, or anything related to death. Sure, there are some "lucky few" who care more about other material things than they do about the essence of the world, life, theirs and others'. For the most part though, people are sympathetic beings who take heed of the well-being of others. Or that's what I like to think. But anyway, the grand majority do care about their own lives at the very least.

So when you bring death into discussion, there is a slight retention in all of us. This accentuates even further when we refer to an actual end of everything, an apocalypse. It is this knowing that not just you will die, but everyone you've ever known, loved, admired, hated and everything that's ever been will perish alongside with you, which hurts to even think about.

There is this obstacle that the human mind finds exceptionally difficult to process, that you and all of us are eventually going to die. Everything we do in this life is, subconsciously related to this idea – it's simply an effort to distract ourselves from acknowledging that someday we are going to die. We might achieve extraordinary things, good or bad. We might be remembered, we might not be. But in the end, we are all heading towards that inevitable point of passing away.

As disappointing as it sounds, this is the way things are. And as of our current understanding of life, nothing can be done to reverse this process. Measures aimed at prolonging or halting it for some definite period of time could, indeed, be taken. Yet in the end, death is inevitable.

"Only two things are certain in life –

death and taxes"

Benjamin Franklin

I don't know what to say about taxes, but I can agree with death. Now I do not mean to get into the meaning of life – that's an entirely different subject, which would require at least another book of this size and there would still be a lot of empty spaces.

Moving back to our theoretical collective death, as humans, as species... incomprehensible as it sounds, it could at some point become a reality. And it is not necessarily the death that is our main concern, but rather when and how it will happen. This complete uncertainty of what's essentially the focal point of unpleasing events makes it even harder to digest.

Throughout the elapsing of time, ever since the beginning of our existence, we have tried to predict, or trick ourselves that we somehow foresee the very end of this existence. In reality, we know that's not the case. But there was most assuredly a point in our lives when we encountered other individuals attempting to do just that, tell us when and how will the finale of all humankind be.

The problem is that there wasn't only that one time. Many, many times has this occurred. And they were never right. You are reading this book now, which means you are not dead. You didn't perish when the world ended at the start of the new millennium. Neither did you in 2012. Neither did you in 2018 when a giant, unknown ninth planet interfered with Earth. And you likely won't in the near future. At least not due to a worldwide catastrophe. But unlikely as it may be, in reality one might occur sometime in the near or distant future. And there is no way to tell for sure.

It is, however, literally vital for us to be able to forecast potential threats to our welfare. Yes, we can calculate the trajectories of asteroids years, decades and even hundreds of years before they get close to us. Yes, we can predict where earthquakes can take place and take measures against them. And yes, we can sign peace treaties to prevent future wars. But if... if somehow a doomsday scenario were to occur, it would do so unexpectedly. We wouldn't realize its exact consequences only once they manifest. And it may be too late.

What makes such dangers ever harder to recognize by the masses and even by those in power are the unstoppable misinformation that is encouraged by news reports. When it comes to fake news... in an area like this, the Armageddon... it can actually become a death sentence for us all. With so many hoaxes like self-proclaimed prophets and Mayan calendar cycles there will be a point, a threshold, when no one will bite it anymore. There will be many fanatics who will believe every apocalyptic announcement, but for the most part, humans will no longer take those for granted.

So what will we do when one of them is, in fact, accurate? How do you think we'll respond? Some may panic, some may stock up on their supplies and prepare for the worst, as it always has always been. But the rest of people... they'll remain calm. They might not even flinch in the face of the greatest danger our species has encountered. All of that because of so much misinformation. And it only takes one correct claim... think about that for a minute...

I was personally very impressed by a movie that was much expected and months after its release it was still thriving on the highest spots of every list. That film is called "Don't look up!" and apart from its world-renowned cast, it also consisted of an exceptionally well-written plot. It is a satire based on a very similar story, two scientists finding out about a planet-killer asteroid, as they are called, which was seemingly going to impact Earth in around six months after the two calculated its trajectory. And in spite of their greatest efforts, no one was believing their warnings and they were regarded as insane. Only when they saw the fireball with their eyes did the people understand the imminent danger. No one had taken any measures to save our species and when they realized their mistake, it was way too late.

It was a really controversial movie, with critics' opinions cancelling each other out. If I were to choose one side though, I would definitely go with the admirers. I found it particularly realistic, dare I say accurate. Of course, there were some out-of-place jokes and I sure hope something like this doesn't happen to us, but seeing how people behaved and the similarity between that society and ours... it just blew my away. The resemblance was staggering.

The biblical cataclysm

There is one prophecy, if you could call it that way, to which people have paid much more attention, and which is far darker than anything we've ever witnessed before. No, I am not talking about some horror scenario from a random book or movie. This isn't random at all. I am referring to the Holy Book of Christianity, the Bible – The book of all books, the Holy Scripture.

This is one of the most widely read books in the history of mankind and it will stay that way for longer to come. Apart from that, other religions' Holy Books also have at least one section dedicated to the end of days (however, I will not express myself on those). But how come that the Scripture, which is meant to bring happiness in everyone's home, ends with the most somber doomsday event imaginable? Before answering this question, I would like to explain a bit why this is a slight issue, and why I have decided to address it.

It is because of the spread of Christianity around the globe that the Bible has gained so much ground in front of other books. People became increasingly more interested in it to a point where almost every Christian individual possesses at least one Bible in their houses. It is true that this religion has fallen through in latest decades, but the Scripture remained of great importance to every believer.

It goes without saying that the Bible immerses you in faith. One key lesson that it teaches us is that we shouldn't be afraid of death, as death is not the end of it. That we must have faith and believe that there is something else other than our mortal nature. But I could almost guarantee that even the most dedicated Christian got a shiver down his back when reading the last book of the Bible – The Apocalypse of John, or The Book of Revelation.

So why does it end this way? Why does the end of the world mark the finale of the most important Christian book? Maybe that's the whole point: that you have gained so much faith whilst reading it that by the time you get to The Book of Revelation, you wouldn't be concerned anymore. Maybe the entire Bible is supposed to prepare you for The End; to make you acknowledge the fact that God is almighty and he could end it all without even lifting a finger (excuse the personification); and that we must only do good things in this world so when The End comes, we'll stand beside Him, and not away from Him[1].

This seems about right, though it further accentuates the idea of an imminent end of everything. Will it be good or bad? We don't know. There is no way of knowing. We can only live with faith and optimism in the hope that it will only be a milestone to our existence and not the passing into nothingness.

The beauty of the Bible consists of the liberty of interpretation. A single verse can be seen in as many ways as there are pairs of eyes, perhaps even more. But due to this liberty, the fact that the human imagination is allowed to almost run wild when reading it isn't as great for this final chapter of it. It might be the lesser evil still, as if you were to know exactly how and when it will happen, you would live your entire life in fear.

An analysis of apocalyptic scenarios

From my perspective, it doesn't matter whether you believe or not, it's everyone choice to believe, not obligation. It doesn't even matter what religion you are, we are all entitled to have faith in something. But what I do think we can all relate to is this sudden epilogue of the human species. We are all humans after all.

To me, one thing is certain regarding it: when the Providence will decide it is time for us to go, we will go, without being able to oppose. But until then, there may be some major threats to the prosperity of our species which we must annihilate and overcome.

With that in mind, I would like to elaborate on the most macabre and yet probably "truest" apocalypse prophecy, by relating my and other insightful people's interpretations of it to the scientifically proven ways the world could end.

"I looked, and there before me was a white horse! Its rider held a bow, and he was given a crown, and he rode out as a conqueror bent on conquest."

Revelation 6:2

"Then another horse came out, a fiery red one. Its rider was given power to take peace from the earth and to make people kill each other. To him was given a large sword."

Revelation 6:4

"When the Lamb opened the third seal, I heard the third living creature say, 'Come!' I looked, and there before me was a black horse! Its rider was holding a pair of scales in his hand."

Revelation 6:5

"I looked, and there before me was a pale horse! Its rider was named Death, and Hades was following close behind him. They were given

power over a fourth of the earth to kill by sword, famine and plague, and by the wild beasts of the earth."

Revelation 6:8

That's how it's said to begin – with the four horsemen, harbingers of The Apocalypse... Naturally, those may not be actual horsemen, and on a closer analysis, we may find out something even darker than that.

Hence, let us begin studying the possible scientific doomsday scenarios that might fit the biblical description. We'll start with the least likely ones and move up on the likelihood scale, according to the four defining categories: highly unlikely, unpredictable but unlikely, probable but avoidable, unpredictable and somewhat likely.

By the end of it, you will hopefully have a broader perspective on the complicated matter that we will discuss. So, without further ado, let's see how the world could end by next Wednesday!

"What the caterpillar calls the end of the world

the master calls a butterfly."

Richard Bach

Four Horsemen of the Apocalypse, an 1887 painting by Viktor Vasnetsov. From left to right are Death, Famine, War, and Conquest, with the Lamb at the top.

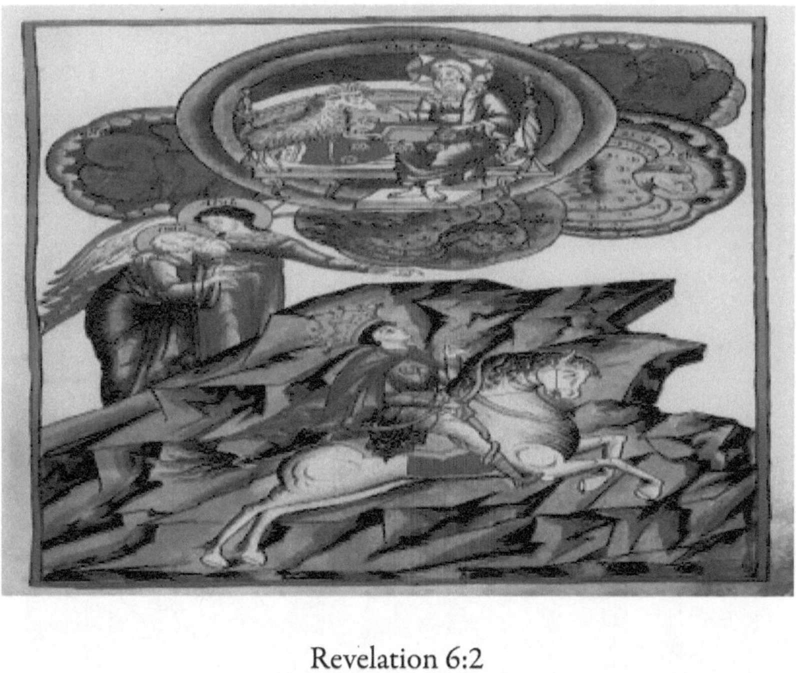

Revelation 6:2

White Rider from Tolkovy Apocalyps, Moscow, 17th century

1. I personally prefer to use “It” when referring to God, and not “He” or “Him” as it suits better with the description of Supreme Being or Supreme Entity. Either way, there is no way of telling what Its nature truly is, so “He” and “Him” may not be the most appropriate pronouns. However, for the sake of clarity, throughout this book I will often use “He”, “Him” and so on, so that no misunderstanding arises.

SECTION I:

Highly unlikely scenarios

About the dangers of outer space – supernovae, gamma ray bursts, black holes and others...

I.1: Universal death

"All good things must come to an end..."

Usually, when we think about the end of the world, the first thing that comes to mind is the termination of life on Earth, or the neutralization of Earth's life sustainability conditions. But the first possibility we'll cover in this book is a disturbance in the grand scheme of things.

The planet we call home, the planet that has harbored all life that we are aware of is but a speck of dust floating in virtually endless space. If life on Earth were to come to an end, business would carry on as usual. The planets would keep on rotating, the stars would keep on shining and the galaxies would look as gorgeous as ever. They wouldn't care. At least at this point in time, we exert close to zero influence on the universe.

As sad as it sounds, our extinction would mean nothing to the rest of cosmos. There was a joke that went something like this:

Earth feels annoyed by human activities so it goes to some other planet friend to ask what can it do to cure this terrible rash called "homo sapiens". The other planet replies – "Well, there are a few things you can do, but its best to just wait for it to pass. I had it too, but now I feel better that ever!"

Obviously, we don't know yet whether there is life on other planets or not, but we'll back to that later. It's important that you got the point.

Then, what if a bigger destruction event took place? Could there be some sort of incident that would significantly impact the other components of the universe, maybe even the entirety of it? The answer is yes, there is, and it's the only scenario that is truly, scientifically inevitable.

So how will it happen? Well, we don't know for sure. But scientists, especially cosmologists[1], have narrowed it down to three big possibilities, namely the Big Rip, the Big Freeze and the Big Crunch. Nevertheless, there are a few concepts that need to be explained in order to fully grasp these scenarios.

Universal inflation

You might have heard of the term "inflation" before, most probably regarding money. What happens there is that with time, the prices of goods are steadily increasing – everything is getting all the more expensive. Alternatively, the issue can be seen as a decline of the purchasing power of any currency, i.e. money losing its value. The term itself comes from the swelling of prices on the market, similar to the increasing volume of an inflating balloon.

Perhaps that isn't the best analogy, but it works extraordinarily well when we are talking about the universe inflating, also known as cosmological inflation. In a nutshell, the inflation is the accelerating expansion of the universe (in its early stages).

It all started at the beginning of it, sometime between 10^-36 and 10^-32 seconds after the Big Bang. This is extremely challenging to imagine. The time intervals are so small, so close to the Planck time[2] that not even the most fundamental particles were formed at that point. It was only space and energy, which drove the inflation, thus that period became known as the inflationary epoch.

And since then the space kept expanding. The balloon is a more mundane example to show what was, is, and will continue to go on. Imagine you drew many dots with a marker on the membrane of the balloon. Assuming that the membrane is incredibly elastic and won't shatter no matter how much air you blow into the balloon, the dots will appear to be getting farther away from each other the more swelled the balloon becomes. Basically, each and every dot will be moving away from all the others and the bigger the balloon, the greater the velocity at which it does so, as the distances increase.

So now let's say that instead those dots we had galaxies or stars or any other body in the universe for that matter. In this case, the space would work as the balloon. However, don't be tricked into believing that the shape of the universe is that of a balloon, neither is the Minkowski space[3]. The important part are the dots, the galaxies in our case. They are all moving away from one another the more the balloon – space – expands.

No single galaxy will stay in place, except from its own perspective. An observer situated in one of the galaxies, any one of them, would see all of the others receding from him, but there would be not one point that his galaxy seems to be receding from (in theory, we know it's the balloon's center that all the dots are moving farther away from, or any set point in the balloon's interior, but as I've said, those don't matter in the cosmological approach – it's not the balloon but its swelling). For this reason, each point would seem to be the center of the expansion, but in reality, there is no such thing. Every point in the universe is moving away from any other.

This is inflation for you. An ever-swelling balloon, where we and everything else in the universe are solely dots situated on its membrane. And this last thing is quite important: that the universe is only getting larger and larger.

In fact, it even looks like the expansion is accelerating. But what drives it? Once again, we have no idea. That's why we called it dark energy.

Dark energy

Dark energy is a hypothetical form of energy. What's special about it is that is exerts negative pressure. Unlike gravity, which attract bodies, dark energy has a repulsive character.

Just like dark matter, we don't know too much about dark energy[4], as it doesn't interact with light, so we are unable to observe it directly. What we can do is perform different gravitational measurements and see how it interacts with gravity (if it doesn't work with light). And that's exactly what scientists did.

The Field Laws of General Relativity, a whole other subject, did somewhat predict the existence of such a form of energy. That was back in 1915 and Einstein called it vacuum energy, the energy of empty space which couldn't be observed directly. Therefore, a cosmological constant was introduced to describe the static nature of the universe[5].

However, once the universe was proven to be expanding, this constant was disregarded from the calculations. It was suited for a non-expanding and non-contracting universe, which meant a constant energy density was homogenously filling the space. The problem arose from the fact that matter and energy are not evenly distributed, so disturbances will inevitably occur. Those disturbances could lead to two models: the open model, and the closed one. We'll move back to the latter a bit later on.

There is also a third model which describes the universe as flat, static as Einstein first assumed. But that is the ideal model, in which what was later found to be dark energy perfectly balances gravity and the energy density is equal to the critical value. None of that happens in reality. The equilibrium is unstable. Any slight disturbance towards the open model tends to make the universe even more open, and the same happens for disturbances towards the closed one.

Back to the open model. The concept of dark energy was experimentally proven around the end of the XX century. It was when a team of astronomers observed supernovae which didn't fit the current model. It was required for a slower expansion to have occurred in the past than it did that day.

Essentially, it was known that the universe was expanding, but since then it is also known that the expansion is accelerating. And it is this form of repulsive energy that drives the expansion.

Intuition leads us to believe that dark energy constantly generates as space expands – many other experiments seem to agree on that. As of today, the universe was demonstrated to be very close to a flat model, however it bends a bit "outwards", guiding it towards the open, ever-expanding model. Finally, how does all this connect with its imminent fate?

The Big Rip

In this first scenario, the dark energy generates at such a high pace, i.e. the universe expanding at such a rate, that gravity poses not even the slightest threat to slow it down. Let's think about what happens with an object that speeds up indefinitely as a quick experiment.

You might be tempted to think that it will move faster and faster and eventually, it will surpass the speed of light[6]. We know that can't be. The expression "speeds up" is a bit confusing here. Any object does, indeed, speed up, its velocity is constantly increasing, but only until a point. A certain threshold that can't be passed, according to, again, the Theory of General Relativity. That threshold is the speed of light, 299 792 458 meters per second do be precise.

So that object will suffer a number of distortions such as time dilation and length contraction, as seen by a stationary observer. Those distortions will aggravate the bigger the velocity of the body, but the velocity itself can't exceed that of light.

Probably the most interesting aspect of the accelerating expansion of the universe is that it doesn't obey the same laws. In theory, it is possible for the expansion of space to happen faster than the speed of light. That is because all the bodies are situated in space, but when space itself is expanding, it's an entirely different story.

What that means is that we know for a fact that objects in the universe can be receding from as at superluminal speeds because, and only because of the expansion of space. We wouldn't be able to make them move at such a speed – well the universe can.

Apart from the fascination that this claim brings, there is also a sad part to the story. As bodies become more and more redshifted[7] (distant) the farther away they are from us, the most remote bodies, the ones at the edge of the observable universe could be receding at velocities above that of light. The fact that you see them as of now is thanks to the finite value of the speed of light – light from the past is only reaching you now so you might be seeing the given object billion years in the past.

Let me emphasize that. You are reading this book which is situated about 30 cm from your eyes. That means that you are seeing it about a billionth of a second in the past – that is 0.3m/299792458(m/s). The Sun is situated almost 150 million kilometers away from us. Consequently, you are perceiving the light that came out of it with 8 minutes of delay (so you see it 8 minutes in the past). Fun fact, if the Sun were to suddenly disappear, you won't know it only after 8 minutes have passed.

The same thing happens for the most distant galaxies – those situated at billions of light years away. Light that reaches you was emitted that many years ago and only now are you able to record it. But if those galaxies are receding faster than the speed of light, not even photons are swift enough to reach you. They never will...

The moment an object situated at about the edge of the observable universe attains a velocity above that limit, light from it will never reach you again. Something very similar to the Sun vanishing all of a sudden. Only this time, you would have to hypothetically wait billions of years of seeing it in the past before realizing it was long gone.

Space will continue to expand, the galaxy will not disappear (unlike the Sun), only our observable universe will be smaller than the actual universe – all of space. At this point, the true Universe is already bigger than the observable one, which means that there were marginal objects we are already unable to see. And we never will. They are lost to us.

From the marginal body's perspective, we would seem to do the same thing. Remember the balloon and that each point seems to be the center of expansion, yet there is no such thing? For a theoretical observer watching the Milky Way[8] from billions of light years away, our galaxy will seem to them as it was at some point in the past, then abruptly fade out into nothingness.

With that in mind, the space between all bodies in the universe will increase at an accelerated rate, depending only on the distance separating them[9]. First, the most distant stars and galaxies will disappear, never to be seen again. Then, the galaxies in our cluster[10]. Then the Milky Way itself. Then our Solar System. Fundamentally, we will be able to see less and less objects because the universe is ripping them apart.

In the end, the Earth will begin to disintegrate, alongside with us, ants, bacteria and all the way down to the smallest particles, the very constituents of the universe. Gravity and the other forces will be overruled by universal expansion and they won't be able to touch anything else... ever again. What a lonely death isn't it? This theory is called the Big Rip.

"The fourth angel sounded his trumpet, and a third of the sun was struck, a third of the moon, and a third of the stars, so that a third of them turned dark. A third of the day was without light, and also a third of the night."

Revelation 8:12

Could this scenario be the end of everything? The Book of Revelation does present a similar possibility. The eighth chapter of the Apocalypse refers to the first of seven angels sounding their trumpets, releasing all sorts of destruction – from fire, to earth, to water, to what seems to refer to the cosmos. In this particular case, the literal interpretation suits surprisingly well the Big Rip evolutionary model of the universe.

Stars would turn dark as they become out of viewing distance. Similarly, the Sun will later on turn dark. The Moon, which is closest to us, will become dark last due to such an accelerated expansion. So, for this reason, the sky would also turn dark as nothing else will be visible anymore.

Of course, this is but one interpretation which suits the theory. In reality, there are many other ways such bodies could "turn dark" before the expansion acts on them, but we'll tackle those in later chapters.

The biblical numbers and symbols such as the fraction marking the turning dark process, "third", could be introduced only to signify that the entire Rip would occur gradually, step by step. Stars will fade out first, then the Sun, and lastly, the Moon.

The Rip amongst smaller bodies such as humans, ants and atoms needn't even be mentioned as the others are more than enough to finish us. So is this the ultimate fate of the universe? Is the Big Rip the one definite scenario? Well, more trumpets are left to be sounded and other theories make it higher up the likelihood list.

The Big Freeze

A similar scenario in which the universe will slowly and inexorably grow darker is referred to as the Big Freeze. This would occur somehow in parallel with the Big Rip.

The Big Freeze theory states that, in an expanding universe, the temperature of everything will slowly but surely decrease asymptotically until it essentially reaches absolute zero temperature. This is the lowest temperature there is, and it can't practically be reached by any object, unless we are talking about the endlessly growing universe (it is the equivalent of 0 Kelvins or -273.15 degrees, in Celsius). It would take an infinite amount of time for such a state to be reached, so the scenario is rather hypothetical. Most of the times, the Big Freeze is regarded to be the same phenomenon as the "heat death" which will later be examined, but we will take it separately just for the fun of seeing what might happen in an eternally expanding universe.

How does this work? After the universe was born, all the energy generated by the Big Bang was spread throughout space, but it was only a finite, however extraordinarily large quantity. At the beginning, the temperature levels were so high to that not even the elementary particles and fundamental forces were able to form. It then steadily decreased so that more complex elements appeared and later on even bigger bodies such as stars, black holes and galaxies. So, as space expanded, the temperature got lower and lower, as the energy dissipated. Nowadays, the universe has a pretty much uniform temperature of around 2.73 Kelvins, slightly above absolute zero.

As it continues to grow larger, the temperature will decrease even further. Intuitively, if the universe continues to do so indefinitely, the temperature will eventually reach absolute zero, and nothing will happen anymore as all particles basically freeze at that point. Another upsetting death... Everything seems nice in theory, but let's dive into the more practical details of this scenario.

In our open model universe, things don't look too bad: stars are shining, planets are revolving and we are having the time of our lives... for now. But no matter what happens to us, for the time being stars will keep forming. Sure, they will live their billion-year lives and sometime in the future they will die as well. Still, they will keep forming for as long as there is enough material – gas – to do so. This is called the stelliferous era and is expected to last about 100 trillion years. For reference, the age of the current universe, a stupendous 13.8 billion years, is around 7000 times smaller than that value, virtually meaningless. Yet there are other much lengthier eras that succeed our current one.

The degenerate era comes after it, and it begins as soon as there is no more stellar material left in the universe. During its first stages, remaining stars will burn out their fuel and turn into mainly white dwarfs, brown dwarfs, neutron stars and black holes. Those periods last much longer than the main sequence life of stars – i.e. when stars are able of or use predominantly hydrogen fusion. White dwarfs are thought to assimilate dark matter, which can prolong their lifetimes a bit. Yet they, too, have an end.

Later in the era, protons are believed to start decaying. Only black holes will resist this stage and they will slowly accumulate all the surrounding matter. Life will already be impossible, but the era will go on until much longer – 10^40 years is one estimate.

With the stars gone, it is time for black holes to reign – the black hole era begins. Things start getting quite boring with only those massive yet alike bodies lying around. John Wheeler, a famous physicist, once said "black holes have no hair", implying the fact that you couldn't tell them apart if it wasn't for their mass, charge and angular momentum, according to general relativity. Stephen Hawking later refuted this claim, stating that black holes do have "soft hair" – a few other characteristics that tell them apart; but most importantly, he also found out that such objects emit some sort of radiation, named after him, which ultimately leads to their death.

Hawking radiation refers to the pairs of particles and anti-particles[11] that randomly generate thanks to spontaneous quantum fluctuations. We'll talk more about those in a bit. What happens is that when such appearances occur at the boundary (event horizon as it's called) of a black hole, only one of them is drawn towards its interior, ejecting the other according to the momentum conservation theorem. This way, black holes disintegrate over time.

But the death sentence attributed to black holes isn't an immediate one. The term ranges from 10^64 years for solar-mass BHs all the way to 10^100 years for galactic-mass BHs. Thereby, the black hole era will last somewhere around 10^100 years[12], vastly larger than the age of the universe.

When this era ends, another, final one starts: the dark era. While black holes generally have surface temperatures merely a billionth Kelvins above absolute zero, the remaining particles of the dark era will endure even lower temperatures. Annihilation events between those will eventually take place and the universe will turn into a void: all that's left is nothingness. Forever.

The heat death alternative

In an expanding universe like ours, a kindred scenario is expected to occur. The heat death of the universe implies an eventual state of thermodynamic equilibrium being reached throughout its entirety – this is generally considered the same as the Big Freeze.

It is rather challenging to fully grasp this idea, so let me break it down for you. A good starting point is the concept of time.

As of our current models which seem to most accurately predict the nature's phenomena, time can be regarded as another dimension of the universe. But unlike the other 3 spatial dimensions that we are aware of[13], time can't exactly be seen. So how do we know to measure it?

The first thing that helped define it was our intuition, our common sense. We remember thing that we've already done, that have already taken place. And we often have proof of it, so they must've been real. But we are only aware of what happens now. So how do we refer to those – we use the notion of past. Analogously, for things that we expect to happen we use the notion of future. And the present is for what we are aware at any moment.

Next, in order to attribute a unit to it, just like we used feet, sticks, arms or others for measuring lengths, we tried to find something which was known to us in order to measure time. The first thing that came to mind were days – the sun comes up and down at similar (though not exactly the same) time intervals. Nowadays, more set intervals are known, thus it is easier for us to determine time much more precisely. Also, all durations can be extremely accurately calculated with the help of the truly refined fundamental unit which we call "seconds".

This notion has existed for quite some time, but only recently has it been so refined, due to the innovative atomic clocks. Those are based on certain changes in the states of atoms and their constituents when specific conditions are applied. Basically, this is the purest form of coordinated events. Everything else can be summarized to that.

What can be extracted from this is that time is brought into existence by such disturbances at the most fundamental levels of the universe. This is how it became related with a key notion of thermodynamics called entropy.

Entropy is most commonly associated with the degree of disorder in any given system. The higher the entropy, the greater the disorder. To better emphasize this property of systems, let's analyze a few familiar examples.

You've probably all heard the following under one form or another: "Clean up your room! It's such a mess in here!". If only you knew that it would be even greater of a mess upon finishing your chore... Don't get me wrong, I'm not trying to excuse you of cleaning your room, but on a molecular level, the statement is quite accurate. Things will be more ordered after you sort them out, but the energy you consume and dissipate while doing so makes the whole system less ordered than before. The same happens when you freeze water. You may know that out of the three usual states of matter (gaseous, liquid and solid), particles are most disordered in gases, then liquids and then solids, where they are the best arranged. So intuitively, when turning water into ice, the body of what previously was water does, indeed, become more ordered, but the energy consumption of the freezer while doing its job made the system as a whole more disordered.

It's as if you threw a stack of paper in the air. When the paper touches the ground, it will no longer be in the ordered for of a stack, it will be pages lying all over the place. This is a more disordered system then the first one. If you then took the pages and burnt them down, all that's left is ashes, which again form a more disordered system than what we had before. You then throw them in the air and the wind will blow them away, who knows where – even more disorder.

As the second law of thermodynamics states, in general, the total entropy of any system does not decrease unless increasing the entropy of some other system. From its definition, the entropy of a system is directly connected with time. Entropy only increases with time! This is important in any thermodynamic system, especially when talking about the entire universe like we do.

As time goes by, the total entropy of the universe increases – it gets all the more disordered. This will happen to the very limit. The universe will evolve to a state of no thermodynamic free energy – i.e. thermodynamic equilibrium. When that state is reached, no other process which further increases entropy will occur. Yet again, nothing will happen anymore. This is the reason why Lord Kelvin disproved the theory of an infinitely old universe.

It's interesting how the heat death scenario doesn't require the universe to reach absolute zero, only the energy must be uniformly distributed throughout space and the equilibrium is reached.

In conclusion, the heat death goes hand in hand with the Big Freeze theory, we just don't know whether the final temperature will be zero or the equilibrium will be attained beforehand. Other than that, entropy works the same in both cases: stars are organized, black holes not so much (they are the most disordered bodies[14]) and the random resulting particles are the very definition of mess. And all while the universe keeps expanding faster and faster tending towards a Big Rip. What a deadly trio!

There is, however, another scenario proposed by cosmologists – not so lonely this time, but rather crowded.

The Big Crunch

And I saw an angel coming down out of heaven, having the key to the Abyss and holding in his hand a great chain.

Revelation 20:1

The aforementioned theories are widely regarded as the correct fate of the universe. In our particular case of a nearly flat, yet open universe, the character of space will only accentuate towards this model as time goes by. In accordance with that, the expansion will continue forever and the universe will meet its end as described by those theories.

If, and there's a big "if", somehow, the expansion halts at one point in time, such that not enough dark energy is generated to drive it, gravity might surpass the repulsive force. If it halts, then gravity will equal this force and we would have a flat, static universe; this is, as I've said, highly unstable, as any disturbance would incline the balance towards either a closed or an open universe. So, for the sake of argument, we will assume that in our "script", gravity wins and we are stuck with a closed universe.

In all its fame, gravity is the weakest of forces[15], decreasing with the square of the distance between bodies. But as the bodies get closer to one another because of it, gravity will become increasingly stronger and the bodies will accelerate even faster towards collision. Imagine this happening to every single body in the universe.

As gravity wins the one battle against dark energy, it will begin to conquer the universe. Just as expansion gets bodies further from each other, a contraction caused by gravity would draw them all closer together. And will do so at an ever-accelerating pace.

The outcome is just the opposite to what we've discussed so far: galaxies will get closer together, they will merge, stellar systems will become increasingly crowded and everything will eventually collide with the other bodies. Gravity will provoke many trajectory disturbances so the entire universe will basically go haywire. It will be literal chaos.

It will take a while until we get there, as was the case with all the other proposed theories, but once the crazy stuff takes place, there won't be much longer until the end of it all. The universe won't be able to handle it anymore and, finally, it will collapse under its own weight, giving rise to another dot-like concentration of energy such as before the Big Bang. So was that the end? Maybe not.

The Big Bounce?

Then I saw "a new heaven and a new earth," for the first heaven and the first earth had passed away, and there was no longer any sea.

He who was seated on the throne said, "I am making everything new!" Then he said, "Write this down, for these words are trustworthy and true."

Revelation 21:1 and 21:5

A follow-up theory is gaining significant ground amongst scientists...

In truth, not much is known about the universe. We are aware of the main forces and the general laws by which it behaves, but there are some spaces left to be filled. Many more gaps which can't be ignored arise when forcing its limits – that is the extremely small and large, the extremely hot and cold, the extremely light and massive... you got the idea.

Basically, when such extremes are reached, the current laws of physics seem to break down. Nothing goes according to plan anymore: the universe stops obeying them and starts behaving like a rebel that does the exact opposite of what you "tell" him to. And there is yet no perfect theorem, law or equation to describe everything there is – all the forces, all the movements of particles and all of their effects.

Quantum mechanics, quantum gravity and string theory are just a few of the theories which are desperately trying to figure out how the universe behaves most fundamentally. If they could accomplish this task, we would know perfectly what to expect for any given conditions. We would be able to predict how the universe behaves, even at those presently mysterious extremes. We would have "the God equation" on our side and the oddities of this strange universe will be at our fingertips.

But as of now, we frankly have no idea what's going on. And probably the most obvious example is the Big Bang: the starting point of the entire universe. In this particular case, we are talking about something incredibly small, yet amazingly dense. A point-like structure which encompassed the total energy of everything, then suddenly inflated and continued expanding until it got to what we now call the universe.

And in spite of its 13.8 billion years of age, the first seconds of its life were arguably the most significant. It was in those couple of seconds that all the universe's particles and known forces came into existence (some of them even died out entirely back then). But what happened before?

There is no clue about the chronology of the Bang before the Planck time (10^-43 seconds) elapsed. It could've been anything there. Many particles and forces were hypothesized for that short period, but there is no available way to test if those are right. It's only speculations. And it's only the tip of the iceberg. What happened before time zero, previous to the effective Bang, is a whole other question.

One theory suggests that there was another universe prior to the Big Bang. Something like ours, only it didn't expand indefinitely, but it collapsed upon itself in some sort of Big Crunch scenario. And then began expanding due to some unknown quantum field anomaly which sparked its inflation.

And while the information we posses is exiguous to make any claims about what was before, we do have the ability of predicting what might come next. Assuming a closed universe, the Big Bounce theory has an interesting continuation to what we might've thought was the end. Although an infinitely old universe was proven impossible, the Big Bounce scenario proposes an infinitely old series of finite age universes. Let me elaborate on that.

With the proposition that our universe came to be from the expansion of an infinitesimal singularity with stupendously high energy density, and that we would get something similar if we were to concentrate everything back again, i.e. via a Big Crunch-like means, the idea of cyclicity emerged. Basically, this theory states that the universe once was a singularity which resulted from the "crunch" of another, previous universe, then expanded, gave rise to what we see today, and sometime in the future will eventually collapse back to a singularity, which will "bang" yet again and so on. This indicates a non-stop cycle of periodic universes which expand and collapse.

How could that be? We saw how a universe expands and then starts coming back down upon itself, but what about the effective birth of another universe from such a singularity? There are still a few things to be said.

I would first like to reemphasize the fact that our knowledge regarding the exact mechanism which triggered the Bang is highly uncertain, practically inexistent one might say. But from scientists' point of view, the most probable cause were the effects of random quantum fluctuations within the singularity. Those could've disturbed the compact space-time continuum and sparked a chain reaction which led to the birth of the entire universe. Space-time then expanded as well and gave rise to "a new heaven and a new earth" – a new everything.

If that was the case, there would be many queer implications: an infinitely long series of universes would necessitate infinitely many different possible universes before and after our current one, as well as endless other identical ones. Such things are even harder to explain as they force the limits of human perception of reality, so we are not going to get too much into those. However, the idea of an end to what we call "home" and the creation of a familiar, yet not necessarily twin one, does seem to resonate with the biblical apocalypse. Maybe the Big Bounce really is the case?![16]

Quantum fluctuations and Boltzmann brains

In our foregoing discussion, quantum fluctuations were blamed for the Big Bangs. But what are those fluctuations really and how could they trigger the biggest explosion in history?

In physics, a quantum fluctuation is defined as a temporary random change in the amount of energy at a point in space. The field values are slightly influenced by the spontaneous generation of particle-antiparticle pairs (you might remember we've met those before, at Hawking Radiation). Over and above the randomness, those pairs do not require any energy to produce, as particles usually annihilate each other rapidly in the process. That's why a decent number of connotations come about.

One of them is the hypothetical Boltzmann brains, named after the Austrian physicist Ludwig Boltzmann. The argument suggests that it is more likely for a single fully-functional brain to spontaneously form in a void than it is for the universe to have emerged as the result of a random fluctuation in a universe in thermal equilibrium, but we are mostly interested in the formation of such a brain. There are two mechanisms by which this could happen.

The first way is through something called "nucleation": non-virtual particles slowly forming a brain-like arrangement. However, this would only be possible to occur in a de Sitter vacuum[17], let alone the probabilistic eternity it takes to form – 10^10^69 years. The other method is through quantum fluctuations which materialize in the Minkowski space. Although it would still take incredibly long (10^10^50 years), the formation of a brain via quantum fluctuations is much more likely in our universe.

So just like particles could potentially rearrange or randomly generate such as to form an operating human brain (until it inevitably deteriorates), basically anything could randomly form. The same thing it thought to happen for Big Bangs.

And just as quantum fluctuations have the potential to create new things, they are also able to cause destruction. You could spontaneously die, the Sun could spontaneously explode, and the whole universe could spontaneously self-destruct. This might seem like the creepiest of scenarios, but in reality, you wouldn't even be able to tell as you die.

However, this last possibility is the lowest of them all. It is virtually impossible for quantum fluctuations to take place so precisely that they kill you or the others. Most of the times, they only vaguely vary the values of fields, without having any effect on the general behavior of systems. There's no reason to worry! A more worrisome possibility comes from another theory scientists cooked up for us.

You know how the universe is so fine-tuned that it allows for the laws of physics to behave as we know them, just perfectly for life and everything else to take form? That could change in but an instant via an unexpected disturbance in the universe's basic state.

The theory focuses on the energy levels of the universe, more specifically, the potentials of its component fields. Because they are all behaving properly as of now and are practically immune to slight disturbances, we know they are in a state of equilibrium. However, that equilibrium might only be local and one field may be more susceptible to change than others, meaning that it could, by some work of randomness, change its local equilibrium state and go after the global one, the level with least energy.

Everything in the universe follows the same law. You might be comfortably sitting in a chair, and be in equilibrium. If you raise one leg, move your hands or lean back, you will probably sit just as comfortably as before, maybe even better. Your equilibrium is not disturbed. But lean out too much and you will fall of the chair, going for the lesser energy state, the floor. You have given way to a new state of equilibrium – a more stable one.

Out of the universe's ruler fields, the Higgs field is particularly unstable, some suggest. It is possible that in its current equilibrium state, it isn't stable, but metastable. It means that a disturbance big enough could entirely disrupt the field in one place, and a difference in its fundamental energy would show up. This is called false vacuum decay (from the false equilibrium state in which the Higgs field might be situated).

It even sounds horrifying... Yet it would entail similar consequences to the previous scenario. Even a very small definitive change in the base energy level of the field would trigger an immense chain reaction that would propagate throughout the universe who knows how fast. Everything it touches would be engulfed into a realm of unknown and nothing would be left the same. Not only it would account for the destruction of the universe, but it would dramatically change its rules. It's all a mystery what would happen next.

But remember, those are only speculations that fit our current understanding. False vacuum decay might be a real thing, or it might not. Even if it is, it would be swift and painless, or it might not reach us in the first place due to the ever-accelerated expansion of the universe. And trust me, there are worse ways to go. This was just the beginning.

Conclusion

All of the above scenarios have been seriously considered by physicists and cosmologists for most of modern history. Yet after decades of debate, scientists haven't reached a consensus as to how exactly the universe will die. The preferred theory out of the ones presented above, however, is either the heat death or the Big Freeze.

It has already been demonstrated that the universe is continuously accelerating its expansion and that it will continue to do so forever. Therefore, the Big Crunch scenario is somewhat ruled out by the observations that were carried out. A spontaneous death of the universe is not a real possibility either, as the probabilistic timescales are much longer than those of other similar events.

We are left with the expansion model outcomes. As mentioned, they could theoretically occur in parallel. It only depends on which one happens first. This is why the Big Rip was also disregarded, as it may take too long for it to happen. Consequently, we will have the satisfaction of a peaceful, silent universal death... maybe too much so. What do you think?

But as inevitable as this is, it still didn't manage to make it any higher up the likelihood list, simply because of the bind-blowing times needed for the scenarios to take place. By and large, each universal death entails time scales orders of magnitude greater than the current age of the universe, which is still incomprehensible for us, mortal beings. And to be honest, it is much more likely for us to go extinct some other way long before...

"If the heat death of the universe is the destination,

it really is all about the journey."

Elon Musk

1. Cosmology is the branch of astrophysics which studies the origin and evolution of the universe.
2. The Planck time is the interval in which light travels one Planck length – the minimal unit of length, which is considered to be a fundamental limit of the universe. Whatever the case, the laws of physics seem to break down at lower distances or time scales.
3. The Minkowski space is a combination between three-dimensional Euclidian space and the dimension of time into a four-dimensional space. In it, the spacetime interval between events is the same in any inertial frame of reference.
4. Dark energy is dominant in our times, with 68% of the universe being consisted of dark energy. Dark matter (another form of matter that unlike baryonic, doesn't interact with light either) makes up 27%, while the "usual", baryonic matter represents only 5%.
5. A static universe is one that neither expands, nor contracts, with the repulsive force being exactly equaled by gravity. It is also spatially and temporally infinite.
6. The speed of light (in vacuum) is the upper limit of speeds in the universe. All electromagnetic waves, massless particles and field propagations (like gravity) travel at this speed in vacuum, yet objects with mass are unable to reach it or surpass it.
7. The electromagnetic spectrum is characterized by the wavelength of electromagnetic radiation. The visible ROYGBIV is only a small part of the spectrum, between 400 and 700 nanometers of wavelength approximately. For reference, the higher the wavelength, the redder the light; the lower the wavelength, the bluer the light. When

the universe expands, wavelengths dilate with it as space dilates, making object look redder. The same happens when an object is receding from you, especially at high speeds, due to something called the Doppler effect. This phenomenon is referred to as "redshift".

8. The Milky Way is the name of our galaxy.
9. This expansion, often called "Hubble expansion", is proportional to the distances to other objects: the further away they are, the faster they recede from us.
10. More galaxies form a cluster, more clusters form superclusters.
11. Every particle has an anti-particle: a counterpart with equal mass, but opposite charge. This gives rise to the notions of matter and anti-matter. When they come in contact, they annihilate each other releasing all of the energy (100% efficiency).
12. 10^100 is a huge number also known as "googol".
13. String theory suggest there are 6 additional spatial dimensions, however they are not perceivable because they are exceptionally small, making up for a total of 9 spatial dimension and 1 temporal dimension. M-theory, developed from multiple variations of the string theory, accounts for another spatial dimension: 10 spatial and 1 temporal.
14. Out of all the bodies, black holes are said to be the most disordered, their entropy being proportional to their surface area (see Bekenstein-Hawking entropy).
15. The four fundamental forces are gravity, the electromagnetic force and the strong and weak nuclear forces. Out of those, gravity is the weakest (don't be confused by the names). In spite of that, it has the advantage of acting over large spatial expanses, every

object deforming the field and acting upon distant bodies.

16. As interesting as it may sound, Big Bounce most assuredly isn't our case. The universe was proven to be expanding and it will do so forever. There is no reason for it to stop expanding and collapse upon itself, making it impossible to have a rebound according to the theory.
17. A de Sitter universe is a cosmological solution to the Einstein field equations of general relativity, named after Willem de Sitter. It neglects ordinary matter and only needs dark energy. It does explain the expansion of the universe however, it is not currently the case. But as more dark energy is being generated and its percentage increases (see note 4), the universe will tend asymptotically towards a de Sitter model in the far future.

I.2: Gamma ray burst

"Then the angel took the censer, filled it with fire from the altar, and hurled it on the earth; and there came peals of thunder, rumblings, flashes of lightning and an earthquake."

Revelation 8:5

"You may fire when ready!" – and then the imperial soldiers pulled the levers, and the power of an entire star was focused on the planets of Alderaan and Scarif. Their inhabitants stood no chance... the last thing they saw was their home planets being blown up to bits by the immense power of the Death Star...

This sounds like something torn from science-fiction. It's actually some scenes from "Star Wars" which wreaked havoc at their time. How could a man-made object be powerful enough to obliterate entire planets? To most people's knowledge there is no such light beam capable of causing so much damage – it's science-fiction after all. But George Lucas knew better.

While we aren't aware of any technology so advanced that it could harness the entire power output of stars and concentrate it all towards a selected target, the universe has a word to say. It goes something like this: "gamma ray bursts".

Abbreviated as GRBs, these bursts are extremely energetic eruptions which, instead of blowing up in every direction, are focused in a small solid angle[1]. In this manner, they are shot out in one direction, both ways in general, wiping out everything in their path. Moreover, their power is not that of a star, but that of a supernova; the only difference being its much smaller area of effect.

To this end, the energy you'd receive from it would be orders of magnitude bigger than that from a supernova. The latter expands in all directions, accounting for a spherical distribution of energy. If, however, that energy was be concentrated in a narrow beam, the effects would be much more devastating, making a Death Star look like a pebble in front of this huge boulder that is the GRB.

The rise of GRBs

It is thought that gamma ray bursts are by far the most pernicious energy releases since the Big Bang itself and a few cosmic events could be responsible for those ultimate weapons the universe has come up with.

The first possibility of GRBs generation comes from those stellar explosions we referred to as supernovae. Briefly put, when a massive star (of at least a couple of solar masses) reaches the end of its life, the last evolutionary stages are consisted of a great explosion. The outer layers of the star are expelled into space, releasing enormous amounts of energy – of the order of 10^44 Joules[2]. This is the equivalent of what the Sun will emit during its 10 billion-year life expectancy, only a supernova will emit it in a matter of days, or even minutes.

At its center, only the remains of the star will be left. In the case of supernovae, we are talking about neutron stars or black holes. And for some of them, the magnificent outburst of energy marking the end of their lives might not roam the space in all directions, but rather most of it will be focused in a particular direction, giving rise to a gamma ray burst which lasts about a minute.

A similar oriented jet of energy can be released in the case of two colliding neutron stars. What happens there is that the two heavy bodies orbit each other at increasing velocities as they are drawn closer and closer together. They disturb the space-time in such a way, that variations in the gravitational field can be felt from far away – energy is first released in the form of gravitational waves.

Then, as the two merge to form a black hole, leftover magnetized gas surrounds the resulted object. The induced rotation engages the magnetic field which leads to energetic particle jets to be funneled from the black hole. Another GRB-generating mechanism takes place when a black hole swallows a neutron star. Both these types of GRBs last only a second or two.

Yet another type, ultra-long GRBs as they are called, may be caused by the collapse of blue supergiants, the formation of magnetars or certain tidal disruption events occurring near black holes[3]. They could last not seconds or minutes, but even hours, which is unusual for GRBs. In one particular case, a burst detected in 2011 lasted over 7 hours.

But why are they so dangerous? What makes GRBs so energetic and long-ranged?

Gamma radiation

Gamma radiation is one type of electromagnetic radiation of the spectrum. If you remember the clarification on redshift we had towards the end of the previous chapter (see note 7), the visible ROYGBIV light ranges from 400 to 700 nanometers in wavelength, but the spectrum extends way beyond the limits of the human eye.

As you may know, light has a dual character, meaning that it can be treated both as particles – photons, and as waves – electromagnetic radiation. And because of this wave nature of light, there is also the notion of wavelength between two consecutive crests or troughs that describes its features. Basically, the smaller the wavelength the more energetic the wave.

When talking about the electromagnetic spectrum, the lower you go in wavelength, the "bluer" the light. Equivalently, the bigger the wavelength the "redder" the light. That is a rudimentary explanation for our quite limited eyes, as the 400 nanometers mark corresponds to violet light and the 700 nanometers mark to red light.

On the one hand, the wavelength can get as large as it wants, there is no true limit to it. Beyond red we have infra-red, microwaves and radio waves, those being the least energetic. On the other hand, it cannot get any lower than zero as it designates a length after all. The closer it gets to zero, there are ultraviolet, X-rays and gamma waves, with the latter carrying the most energy. In fact, a single gamma photon is more energetic than millions of visible photons combined. This is why gamma rays represent a form of ionizing radiation.

This indicates that they are more than capable of knocking electrons out of their orbits, breaking atomic structures and even chemical bonds in the process. You know how it isn't indicated to spend too much time in the sun as it can badly influence your skin, even leading to skin cancer in extreme cases? Well, that's the power of UV light for you. Now imagine something thousands of times stronger than that – those gamma rays.

The sun also emits them, but only in tiny quantities. The space, however, isn't so kind. High-energy radiation roams the universe as leftovers from the Big Bang. So how come we are not affected by it?

For starters, the Sun's magnetic field acts like a protective mantle surrounding the Solar System. That is called the heliosphere. And as good as it works against letting ionizing galactic radiation reach our Solar System, it doesn't affect uncharged gamma rays. Still, we don't hear too often about gamma rays posing a threat to us.

This is because we benefit of another layer of protection, the Earth's ozone layer. While our planet's magnetic field is not nearly as encompassing as the Sun's, the ozone layer does a pretty good job at reflecting incoming gamma radiation. Trivial amounts are, indeed, able to pass through, but for the most part, it's safe to say that they don't pose any danger. On other planets such as Mars, which don't have an atmosphere as kind as ours, radiation levels rise to way higher levels. Even the Moon is significantly more irradiated than Earth, which is why astronauts need to wear ultra-protective suits. That's a reason to thank our planet!

A chill down the spine

If this is the case – that Earth's ozone layer reflects incoming gamma radiation so it doesn't reach us – are GRBs really a cause for concern? They might be...

You see, dangerous radiation is omnipresent in the universe. We are living in a hostile environment, a universal jungle if you wish, which is constantly trying to kill us. Fortunately for us, we happen to thrive in a life-sustaining habitat with conditions that allow us to perform our activities as usual, without too much interference from outer space. But did you seriously think that the universe gives up that easily? Oh no, it sure doesn't!

While common cosmic radiation is feeble and can't compete with our ozone layer, the universe's deadliest weapon definitely can. Gamma ray bursts are terrifying for two reasons: their utter power and their long-range action.

The immense amount of energy released has already been covered. But because the entire energy is focused in one direction, the radiation density is much greater. Think of it as a water gun, only much bigger and much deadlier: if the water was thrown from a bucket, it wouldn't hurt you, but when the jet is concentrated, you will undoubtedly fell it pinching you. For reference, in a couple of seconds, which is their average spam, a GRB released approximately as much energy as "100 trillion nuclear bombs. Going off every second. For 100 billion years."(astronomer Michelle Thaller)

Sadly, not even the ozone layer stands a chance when opposing such a beast. Its shielding will be irreparably deteriorated, if not completely extirpated by the fingerprint of a gamma ray burst. This would lead to either direct or gradual destruction caused by the inability of what's left of the ozone layer to protect us from Solar radiation.

But the power of a GRB that reaches Earth depends on its distance from us – it decreases with distance. That is because the area of the focusing cone of energy is wider spread out, while the angle at its origin, however small, remains the same. Therefore, radiation density will be smaller and the jet might not be as harmful as detailed before.

Even so, it will be able to reach us. GRBs have the potential of travelling millions and billions of light-years across space and still transmit great quantities of energy. They are essentially "cosmic snipers" as some call them: once their gun is pointed towards you, it will surely hit (but maybe not as hard).

We know this because it has happened to us in the past. In 2008, a GRB impacted us directly and satellites were able to record it. But how come we are not dead? The GRB was determined to have come from 6 billion light-years away, from somewhere on the far side of the universe. Its origin was so distant that by the time it got to us, it has almost entirely lost its punch. We escaped unharmed. That time...

Close calls

In 2008, lucky star-gazers were able to see the bright light of the GRB. To them it looked like a new star shortly came to be before immediately fading out. That was because of the GRB's afterglow, as it also emitted visible light besides gamma radiation[4]. And it came from over 6 billion light-years away! Imagine a glow as bright as that of a star, but instead of 10 maybe 100 light-years away, it came from the outskirts of the universe.

That was one GRB detection, a bright one. In fact, many more bursts like that are observed almost daily, but maybe not that intense. There is still one treat that is common among all of them: they are all coming from other galaxies. Out of the hundred billion estimated galaxies in the observable universe, it is expected for GRBs to occur in some of them, even though it takes around one million years on average for one to form in a particular galaxy. And only a small percentage of those are directed towards Earth so to be recorded.

Back in 2004, a short GRB which lasted only a fifth of a second was observed from outside our galaxy. This was caused by a phenomenon called starquake within a magnetar. A starquake, similar to an earthquake, is produced by internal movements. But in contrast with an earthquake, a starquake doesn't alter the stellar surface. It's the magnetic field which is altered, making a hot jet of gamma radiation get ejected out of the magnetar.

The burst was powerful and, gladly, it came from far away. But what if it came not from outside of our galaxy, but from our own? What if it came from our stellar neighborhood?

Conceivably, the effects would be much worse. We can only wonder what would happen if it were for a GRB to target us from closer by. Actually, we might have a clue. Though there are only speculations, some say that a GRB might have already caused a mass extinction event.

Nearly 450 million years ago, back to around the end of the Ordovician period[5], a world-wide cataclysm wiped out 85% of all life on Earth – actually all marine at the time. One possibility is that the ozone layer was damaged by such a cosmic explosion, leading to increased temperatures everywhere on the globe and the peril of sea creatures. This was the second biggest extinction event in history, after the Permian extinction[6].

Maybe one could happen again...

Could it?

Only time will tell.

There is no way to predict when a GRB may take place. Even less so whether it will hit us or not. And for most of the cases, we don't even know if there is any probability at all for a GRB to occur in the future.

Supernovae are unpredictable – you don't know for sure when one will happen. We can only estimate which stars might go supernova in the near future. But that term, too, is vague when referring to the cosmos. It could happen in 1000 years, 10 000 years or while you are reading this. It could've happened already. Still, there is no way to tell because light (information) hasn't reached us yet.

Same goes for magnetars. No idea when a starquake might happen, we can only hope that it won't. And there are magnetars nearby so it's quite unsettling to think about it.

The story does sound a bit different when observing potentially hazardous binary systems of neutron stars or NS-BH systems. We can calculate quite accurately how long we have left before the eventual collision, but there is still no way to fully predict the direction of a possible GRB arising from it.

The only time we'll know for sure if the GRB was a hit or a miss will be the moment light from it finally reaches us. And in the case of the sniper hitting its target, planet Earth, it would be too late. The deed is done and there is nothing we can do to change that.

To us, a world-destroyer GRB would be quite a swift death. We wouldn't know when it will come. We would literally blink and we'd be gone. Not so bad, is it? But if a burst like this wouldn't eradicate us immediately, things won't be as pink. We would see the burst, but it will take a while for the Sun to make our planet's conditions no longer suitable for sustaining complex life. Our demise will be slow and harsh. And there are quite a few not so distant candidates for such a doomsday scenario.

As of today, no GRBs have been observed within the Milky Way. The closest analogs for the formation of long GRBs are perhaps luminous blue variables or Wolf-Rayet class stars (the latter are extremely hot and massive ones, which have shed most or all of their hydrogen envelope). Eta Carinae, WR 104 and Apep are a few of the possible past or future GRB progenitors situated in our vicinity. Still, it is uncertain whether those stars have the appropriate characteristics to give birth to such a burst.

Conclusion

Gamma ray bursts are by far the most dangerous weapons the universe has in its arsenal. But unlike a normal weapon, those cosmic phasers aren't planned to go off. They may do so at any time, without the slightest of warnings. So, should we be worried?

Yet again, not really. Notwithstanding the GRBs horrifying potential for destruction, the rate at which they trigger is quite tenuous. Yes, satellites do record multiple such explosions a year, but each and every one of them came from outside of our galaxy. And yes, only some of them reach us (those pointed at us), so there is no way of knowing exactly how many activate every second. But all of them have been harmless so far: the extragalactic snipers' bullets cannot hit us at full punch. The few worrisome candidates inside of our galaxy may prove harmless, too.

With those in mind, there is no way gamma ray bursts made it higher up on the danger list of our apocalyptic scenarios. Still, the universe is full of dangers which we can't ignore, and one of its devastating weapons, just like a Death Star, "may fire when ready" at any point...

"Gamma rays are the sort of radiation you should avoid. Want proof?

Just remember how the comic strip character «The Hulk»

became big, green and ugly."

Neil deGrasse Tyson

1. A solid angle refers to a 3D way of measuring angles. It works as the angle at the vertex of a circular cone, and it's calculated as 2*pi*(1-cos x), where x is the usual, half-angle described by the cone. The unit of measurement is the steradian, with an entire sphere describing 4*pi steradians. Small solid angles like the one we were talking about are somewhere around or below 10^-4 steradians.
2. Energy is measured in Joules. For reference a full tank of fuel uses around 2*10^9 Joules for a usual car. 1.5*10^22 Joules reach our planet daily from the Sun. A supernova accounts for an energy over ten billion trillion times bigger than that.
3. Magnetars are neutron stars with extremely intense magnetic fields. The magnetic field decay powers high-energy emissions, such as GRBs. Tidal disruption events occur close to black holes, as those massive bodies stretch and swallow stellar material. An accretion disk is usually formed around it, leading to temporary flares of electromagnetic radiation, as the disk is consumed by the black hole.
4. Collisions between the ejected material and interstellar gas on its way towards us give rise to longer wavelength emissions, some reaching the visible spectrum.
5. The Ordovician is a geologic period, the second of the Paleozoic era. During this period, a significant biodiversity increase event took place, with the first arthropods making their way to land, as well as the first plants. A major extinction happened at the end of this period, marking the beginning of the next one, the Silurian. Even though most of life was eradicated, it continued to flourish and develop both on land and at sea.

6. The Permian extinction determined the boundary between the Paleozoic and the Mesozoic. It occurred 252 million years ago and is considered the worst extinction event that has happened so far. 95% of marine species and 75% of land species perished that time, as water temperatures rose above 40 degrees Celsius, caused by the destruction of the ozone layer as a result of volcanic eruptions. The entire event is thought to be the work of an asteroid which impacted Earth.

I.3: Black holes and supernovae

"The heavens receded like a scroll being rolled up, and every mountain

and island was removed from its place."

Revelation 6:14

Out of every fragment of the final book in the Bible, I found the one above particularly confusing. How could the heavens recede like a scroll? And how could chunks of land of those dimensions simply be lifted up and removed from their place?

Obviously, the text is undoubtedly symbolic. Words receive secondary meaning, become more of a metaphor than an actual prophecy. It would be foolish to think that the skies will be rolled up like a papyrus and the Earth shattered to pieces because of it.

However, on a closer look, a disturbing correlation with science emerges. Yes, this seemingly unreal biblical scenario might actually signify a very real possibility for the end of days. What? You really thought that the previous chapters summarized all of the universe's plan to get rid of us? You were so wrong... the plan is much longer than that. It's still got many cards up its sleeve.

One of them has been troubling scientists for centuries and from now on, it may even trouble you. No, but seriously, this has been a true challenge for astrophysicists ever since they began understanding the cosmos. Stars were nice and all, galaxies spectacular, nebulae even more so, but then there are those dark, ugly monsters that devour everything in their path – black holes.

Most of the universe's dangers come in such a way that it is nearly impossible for us to analyze them and know what to expect. Black holes are no exception. They force the limits of space so much that space itself can't cope anymore, and neither do the laws of physics when it comes to these extremes. Even so, throughout the history of science we did manage to decipher some of black holes' mysteries. And while these make up surprisingly useful information for the study of cosmology as well as quantum physics, they also give way to a wide range of jeopardies for life and not only.

The formation of black holes

Black holes are, by no means, easy to understand. Their formation is a complex process in itself, let alone the puzzling result. Still, I will attempt to put it simpler so that the following phenomena won't be as difficult to grasp.

Usually, it all begins when a massive star reaches the end of its life cycle. After millions or billions of years on the main sequence[1], it's about time our star retired and transformed to something dimmer. It might turn into a neutron star, a dwarf, but if the star is especially heavy (several times heavier than the Sun), it will mark its end with a light spectacle before becoming dark forever...

This happens as the radiative pressure inside the star no longer counteracts the gravity pulling on its layers. As it falls short, the outer layers start collapsing upon the inner ones, making the star ever denser until the very foundation of its atoms begins to break down. Atoms smash into each other, giving birth to white dwarfs. If the gravity is stronger, then electrons smash into photons, resulting in neutrons, hence neutron stars. And if the mass is even bigger and the gravity heftier, then it breaks down to the level of quarks[2], leading to black holes or the hypothetical quark stars.

That's about stellar black holes, the ones which result from collapsing stars and which have masses of a couple of times our Sun's. But not all black holes go through the same process.

Despite bearing resemblance with one another, they differ quite a lot in a few aspects, as mentioned before (spin, charge, mass). But probably the most fascinating of those is their varying masses.

Let's take galactic black holes. No, those are not formed from entire galaxies gone crazy. Instead they are generally situated at the center of galaxies. And they are reputed to be some of the most massive and compact objects in existence.

In contrast with their stellar counterparts, these black holes are called supermassive for a reason: they are enormous. They span from millions to billions of solar masses. Our Milky Way galaxy has such a monster at its core, Sagittarius A* as it's called, with mass of around 1.5 million solar masses. And our galaxy is a small one. Andromeda, our closest neighbor, has a similar black hole at the center, which is bigger than ours.

In reality, we don't know for sure that those are black holes at the centers of galaxies, but observations and galactic trends[2] have led us to believe that supermassive black holes (SMBHs) are the most reasonable explanation. And because we believe so for the galaxies in our vicinity, we assume that every galaxy in the universe is alike in that way – this is called the Copernican principle, or rather a derivation of it. We are not special in this huge spatial expanse. Why should only we be gifted with galactic monsters?

So how did those form? There is no unanimously accepted theory regarding the process behind SMBHs. One model suggests that they originated in the early universe, when giant clouds of gas or the predecessors of today's stars, something called quasi-stars, gave birth to larger black holes than usual. But it doesn't even come near to the million solar masses milestone for SMBHs. Yet due to their early formation, the primordial material in the young universe sufficed them. They could've just kept assimilating more and more mass (baryonic or even dark matter) to the point when they became supermassive. Galaxies eventually formed around those accumulations of mass, hereby our familiar universe.

Others speculate that the Big Bounce theory is better able to explain the existence of SMBHs, as it's thought that they were formed before our universe in the everlasting series of bounces. But this theory is more of a funny imagination exercise and not so much of a palpable model.

There is also an upper limit to which SMBHs can grow. They could potentially merge together and form what's known as ultramassive black holes, but they have a limit of a few tens of billions solar masses. Yes, they could ingest even more mass, especially from swallowing other black holes, but their general surrounding accretion disks begin to slow down from a point and coalesce into stars instead.

On the other end of the spectrum, we have small black holes which could be generated on a quantum scale or even created in laboratories. The only thing required is enough mass and a means of packing it in a sufficiently small volume.

The issue with those is that Hawking radiation will severely impact any of our creation attempts. While it takes trillions upon trillions of years for a regular black hole to evaporate, it would take fractions of a second before an outrageously tiny black hole like these ones passes away. Over and above that, all of its mass will be converted to pure energy.

Now this might not seem like much, but let's have a look at some examples. Maybe there is a mad scientist out there who wants to transform his annoying cat into a black hole. At first, he wouldn't think too much of it, until it begins to dawn to him the incredibly minute size such a black hole asks for. He would have to converge all of the cat's mass (about 4 kilograms) into a sphere[3] with radius smaller than 6*10^-27 meters. Assuming that he somehow pulls off this paramount task, he wouldn't even have time to blink before the entire laboratory goes boom!

In the slightest fraction of a second, the man-made object smaller than the most fundamental particles by orders of magnitude will convert its mass into energy without any losses, generating a great explosion. For reference, the energy released is enough to power up an entire country for a full year, that's how destructive it can be.

If the scientists reflected on the issue beforehand and tried to do something similar which might not evaporate instantly, he might try the experiment with a plane. Hundreds of tons are clearly better than 4 kilograms, aren't they? Not so much really.

The resulting black hole would still have dimensions of the order of 10^-24 meters, still amazingly tiny. This time though, it would take a bit longer for it to evaporate, but this cooldown period wouldn't last more than a second either. However, when it does go off, it will cause an explosion tantamount to 5 million megatons of TNT – about 90 000 times more energetic than the largest nuclear device ever built. It would, quite literally, blow up our planet and liquidate all life on it.

The monster's nest

So black holes come in numerous sizes and swallow everything in their path. Yet it might surprise you how simple (at least apparently) their structure is.

Creating a black hole is a daunting task for us, humans; and a dangerous one I might add. But when it comes to the universe, it doesn't even "lift a finger" and voilà... a new one is formed. That's thanks to the simplicity of the principle behind them.

In order to get a black hole, a simple criterion must be followed: there needs to be enough mass in a sufficiently compact volume. Should this happen, the gravity inside it would be so strong that not even the swiftest of universe's components, light, would be able to escape it.

Let's try another approach. According to the theory of relativity, the space-time foam is deformed by every object with mass. Whether it is an atom, a coin, a planet or a black hole, every normal object distorts the fabric of cosmos by a certain amount. You can visualize this as an elastic membrane – your bed sheet for example.

First, is it flat, like if you stretched the membrane across a frame and then fixed its ends so it stays that way. Then, you add a body. Any body will do, but we'll consider some metallic balls for clarity. Those have mass – not too much as they aren't particularly heavy, but they account for a few grams. You might see that the membrane begins to show a few signs of depression. If you then added bigger objects to the equation, let's say a cabbage, the depression will deepen. And you will see the balls falling towards the cabbage. You can even remove the marbles and see how the cabbage dents the membrane by itself. Lastly, you decide to see what a bowling ball does to the stretched material, which presumably doesn't break or detach from its supports. You'll notice a true hollow in the membrane, making it nearly touch the ground.

Another question arises: why does the cabbage (with a volume just as big as the bowling ball's) have so little impact on the material when compared to the last object? Intuitively, it's because of the difference in mass. You see, it's not the volume that determines the depression. It's the mass. We could have a giant balloon filled with air and the bowling ball would still affect the membrane way more significantly.

You might be tempted to think that the density of the object is responsible for the deformation. However, this is not true. In our elastic membrane case, the Earth's attractive force acting on the object is, in turn, also acting on the material, thus deforming it. This force is independent of density. The mass is what matters here.

Same goes for black holes, we'll see in a second why. But to prove this point, a black hole of the size of Earth would have a density of 4*10^12 km/m^3, which is monumental! In opposition, supermassive black holes have much lower densities, however still big by earthly standards. The one at the center of our galaxy has just 10^6 km/m^3. A hypothetical, ultramassive one the size of our Solar System, spanning all the way to the orbit of Neptune, need only be about 1 km/m^3 dense – as if that space was filled with air! Yes, air. That light of a substance.

Now moving back to the membrane analogy of space-time: the fabric of space behaves pretty much like our proposed experiment. Sometimes, such a 2D illustration of the four-dimensional space is used even by scientists to easily understand or present the distortion of it.

In the real world, we see that objects behave normally and don't make troughs in space, but that's what our 3D adapted eyes are capable of seeing in the 2D projections of objects on our retina[4]. In the actual 4D space-time of the universe, things are a bit different.

We might not perceive it, but every object with mass accounts for a small deformation in this fabric. Effects of it can be seen in the strength of gravitational fields, which can be described as a matter of the object's mass at any point in space, as well as in distortion of light paths – we'll get back to those later.

Gravity is important for now. It is because of extreme gravity that these objects are called black holes. Neutron stars are heavy and extremely dense, but their gravitational field is still no match for that of black holes. We can see neutron stars. Black holes, on the other hand, we can't. That's because they are black, and no radiation is emitted for us to receive. That, in turn, is due to the inability of light to even escape it.

In our membrane analogy, the slope was nearly flat far away from the center object and got steeper as we approached it. You can think of it the same way. Light, with the upper limit to speeds, does move slower because of the considerable slope, but as it gets steeper, there is one point where it just can't manage to go fast enough and stays there. It can't escape it. That threshold point is named "the event horizon". If we move even closer, light will not stay in place but be dragged down into the potential well[5].

A more mundane example is that of a waterfall. You (light) swim along the river, in the way of the current. It's weak at first and you could easily go back. You can feel it get stronger as you advance, but your swimming skills still overcome it. As you get to the waterfall, you realize your mistake. You begin falling, with no way to swim back up again, despite your athletic constitution. That is the point of no return, the event horizon, from which point on there is no way back. Not even for light.

This is how the nest-like denivelations of black holes attract other stuff. The current – gravity – increases the closer we are to it, making us accelerate and travel at ever higher speeds, until we reach the cascade – the event horizon – from which there is no way back.

Falling into a black hole

Light is the universe's mightiest athlete. But not even light is able to conquer the universe's steepest slide – black holes. It just gets absorbed into them like any other not so mighty particle. But what happens when bigger bodies meet the same fate?

For this mind exercise, suppose you are somewhere stranded in outer space, in the vicinity of such a dark beast, where no other body can be seen (except maybe your rocket ship) and where nothing besides the black hole exerts gravitational pull – you and your rocket don't make any difference when standing near it, even though you, too, deform space by a bit. You can't see many things, except for distant stars and something weird: a portion of space where no stars can be seen. It is spherical of shape and stars seem to get crowded towards its edges. You remember that physics lesson in high school when you studied black holes and it begins to dawn to you that the danger which awaits you is nothing but one of those.

You realize there is no way to escape it, now that your spaceship is broken. Basically, you are stuck between a rock and a hard place – either you try and fix the rocket even though you know it will very soon be eaten by the hound, or livestream your getting sucked into the black hole to prove the others how is it like to die in such a way.

You decide to accept your fate and dedicate your last moments to the advancement of science, by filming the event with your cockpit camera. Upon stepping outside your shuttle, you notice the black hole has got closer since you last glanced at it. You go feet first so that you can wave to your viewers goodbye. So, what happens next?

To your viewers, it will be quite a calm experience. They will see you recede faster and faster (even though the camera is moving as well) and then simply fade away. Once you reach the event horizon, no even light is able to escape, so the camera won't be able to record you anymore. To the viewers, however, it will appear like you are stuck at the edge of the black hole for a while, and then slowly fade as light gets sucked in. By that point, you will already be inside the black hole.

For you, the experience won't be as pleasant. You conclude that diving feet first wasn't the brightest idea. As gravity increases exponentially as you get closer to the black hole, the difference in pull between your head and feet will be significant. Consequently, you will slowly expand, your column will be so stretched that it breaks and your body will eventually split in half. And then again, and again, until all that's left are your very constituent particles. This process is called "spaghettification". What a delicious death!

You will likely be dead before figure out what's really going on. Your viewers might be disappointed at first, as nothing interesting happened besides your magical fading. As the camera turns away, it gets sucked inside as well. The problem is that the viewers won't know what happens as the connection fails. The video can no longer stream – electromagnetic waves get trapped inside.

However, on the film of the camera a few other things can be seen. As you look away from the black hole, you will see the cosmos like normal. But as you cross the event horizon, something even stranger than your fading takes place. Stars begin to move, explosions occur, galaxies deform. What can this be? It's actually the universe in reverse!

Light that got trapped there a long time ago is able to reach the falling camera. The lonely device will have the chance to see the entire history of the universe backwards before ultimately getting torn to pieces.

Another disappointment

You passed away somewhat peacefully, accomplished and convinced that your video will at some point make it to the rest of our civilization. That's good. If only you lived longer to understand that your last efforts were pointless...

The story takes a sad turn (as if it wasn't sad enough): humans don't receive your signals. Not now. Not 1000 years later. Not ever. This is because the signal didn't reach them in the first place. Assuming the waves were directed towards Earth and you started the transmission while still not too close to the event horizon, your signal must've reached the others. Why didn't it? Here is where Einstein's theory of relativity speaks its mind.

As explained before, space-time can be thought of as an elastic membrane, which is distorted depending on the mass of bodies. For black holes, this distortion is very deep – more like a hollow than a dent.

Now let's take a like beam, made of photons, which pass nearby. In our real-life analogy with the bowling ball in the middle, marbles can be thrown not directly at it, but close to it. It can be seen that they neither fall straight away, nor do they keep the initial linear trajectory. Instead, they go round the ball, passing through the slightly deformed space and then continue their way but on a different trajectory. Basically, they take a turn when inside the ball's dent and proceed at a different angle than the original trajectory.

Light manifests the same way. Once it passes a considerable gravitational field, it takes a significant turn and then continues its movement on a different route. This phenomenon is regarded as "gravitational lensing", because the field acts like a converging lens. A renowned gravitational lensing event was observed in 1985, when a massive galaxy created a gravitational field so strong that a quasar situated directly behind it (thus it shouldn't have been seen) managed to transmit us its light not once, but four times. Instead of not seeing it at all, we saw four images of the quasar positioned in a cross – "Einstein's cross".

The cross configuration was a bit peculiar, due to the quasar being off-center (behind the galaxy, but not on the exact same line of sight) and the elliptical shape of the galaxy. Generally, gravitational lensing is in the form of Einstein rings. Light would be coming from all directions and we would see it all converged like a circle around the massive object. Same thing happens for planets and stars, though not as significant a change can be seen. Black holes are obviously no exception.

This is why, in your desperate aim to direct the signal towards where you knew Earth was you forgot to consider the gravitational lensing effect of the black hole. Light, therefore, takes a different course, never making it to the rest of mankind. What a pity...

"So Einstein was wrong when he said, «God does not play dice».

Consideration of black holes suggests, not only that God does play dice,

but that he sometimes confuses us by throwing them where they can't be seen."

Stephen Hawking

Singularities and the information paradox

After your body and your camera and your shuttle get torn apart, their most fundamental components will still continue their descent into the belly of the hound. But for how long?

To every dent or hollow, every potential well for that matter, there is a lowermost point. A point where potential energy[6] is minimal and which denotes a stable equilibrium state[7]. Those terms are a bit more technical and I will detail them in the notes.

But basically, if you have some sort of deformed terrain, there will always be a lowermost point towards which any given body will tend. In the very symmetric case of the space-time foam, the deformation has its center point at the lowest position – again, visualize it as our elastic membrane (in 3D, not 4D).

For this reason, any object in space, where lots of bodies create dents, will lean towards the nearest point of stable equilibrium. When close to a black hole, that point will be the center of the hollow governed by the mass of the black hole.

But that would happen for stars and planets and asteroid, too. Nothing special until now. There is one key difference, nonetheless. For black holes in particular, the entirety of their mass is believed to be concentrated at that center point, and nowhere else inside of it. All the mass in one infinitesimal point of near infinite (if not infinite) density, called the singularity, situated at the very center of the black hole. The stomach of the beast.

That's where everything ends up. Unfortunately, we can't see what happens down there and also the laws of physics break down at such an extreme. To put it plainly, we have no idea what is going on at the singularity.

The dimensions of the singularity are clearly very reduced, just like the Big Bang singularity. But this point of enormous mass and minuscule size does, presumably, have a small dimension of a Planck length (we discussed about those in the notes of I.1). And all incoming food deposits there.

Other speculations go something like this: black holes usually form from rotating bodies such as stars or gas clouds. Thus, because of the conservation of the moment of inertia[8], black holes are smaller and spin faster. Every black hole appears to do so. And it must do so at any scale. This is why, some scientists imagine that the singularity is not actually a point, but rather a one-dimensional ring, which spins extraordinarily fast.

But no matter the case, it is commonly-agreed that the mass is sedimented at the singularity. So what's the problem with that? Something called the information paradox comes in.

Imagine you have a book – another one, not this one and not the one we broke apart to demonstrate entropy. You get sick of reading and throw it outside the window. It gets dirty, maybe wet, but you can still make out what it says, its all there. In anger, you rip its pages (again). Still, the information is still there. You then throw it in the fireplace and burn it. All that's left is ashes, but it's the same book – except it turned into something else. Essentially, if you had a sophisticated device capable of rearranging the atoms of the resulted ashes, you could get the initial book. That means that the information is still there. It's not lost, only deteriorated.

Physics claims that, in our universe, information can never be lost. It's always there, under different configurations. Well... that's until it breaks down, which is the case for black holes.

Stephen Hawking proved that, information can be lost inside black holes without violating any outside laws. This is known as the information paradox. Intuitively, this can't be. But on a closer analysis, information is, indeed, unable to escape black holes, we saw earlier why. But is it really lost? Yet again, opinions vary.

Optimists suspect this isn't the case. Either the black hole spits out the previously ingested information through Hawking radiation over the course of its life, or it transfers it to a whole other place. The former possibility states that, just like it sounds, black holes act like a temporary prison chamber for information – once it serves its sentence (i.e. the black hole slowly evaporates), it will be free again and not truly lost. The latter isn't quite as optimistic a scenario, meaning that information doesn't make it out the same way it got in. Instead, it is supposed to funnel down towards the singularity and then somehow be transferred to another place, not necessarily even in our universe.

Things start getting bizarre in this presumption. Our four-dimensional space doesn't, in fact, look like that plane membrane model we got used to. It is quite weird of shape, but we can't accurately visualize it because of our limitation. For this reason, it allows for deformations of different places in the universe to somehow coalesce, however far they might be. It only depends on the shape of it (this time we are not talking about the open, closed or flat models of the universe, but rather the exact shape of space-time). This open a completely new horizon of expectations. Things such as worm holes between distant points, on opposite sides of the universe, may, in fact, be possible. We only need this kind of gravitational gateways.

If this is true, then what falls into the black hole might spawn somewhere else, in another gravitational dent, where it could somehow be able to escape. Generally, this is not what would happen. As the black hole creates a much stronger gravitational field, information would still be trapped around its singularity.

There is another possibility, though. What if on the other end is something with negative mass? Something which doesn't generate a dent in space-time, but a hill – this way objects will be pushed away and not dragged towards it. This is a theoretical type of matter, called exotic matter, whose negative mass would impose a repulsive gravitational force![9]

Such objects of negative mass are labelled white holes and were hypothesized as the exact opposite of black holes – they don't absorb all light, they emit it all. Light would be unable to get into them; therefore, they would appear bright. There is no evidence of their existence, but they would change everything if they were true.

Those might be situated in opposition with black holes, on the other side of them. As information is sucked up a black hole, it would pass from its singularity to that of the white hole, and then inevitably get spit out of it at some other point in space.

Theories suggest that such white holes could pe thought of as gateways to other universes and that an entire network of black holes-white holes could be spanning the multiverse[10]. I repeat myself, only theories for now, but who knows? The following fragment could be regarded as a symbol for such an interpretation of the information paradox, too:

Then I saw "a new heaven and a new earth," for the first heaven and the first earth had passed away, and there was no longer any sea.

Revelation 21:1

Pessimists have a much more dissatisfying view on the information issue. They fully abandon the idea that information is, in the end, conserved, and think of black holes not as prison sentences, but as death sentences for it.

They claim that black holes don't ever let information escape, not even through Hawking radiation: it is absorbed and forever lost in the singularity, even after the black hole evaporates. This way, as the universe progresses and the black hole era begins, those insatiable monsters eat each other until only one giant black hole to rule them all remains. All of the universe's information is thrown away and deleted with it, just like you delete a file on your computer. Sad, yes, but they might be right after all...

Does a black hole want to eat me?

Oh, be sure it does! Desperately so! All of them do! But luckily for you, they can't.

Black holes are the least pretentious bodies in the universe. You might not fancy a particular food, so you avoid eating it. In contrast, black holes fancy everything. They eat all the food you serve them, no matter its type, its quality, not even its expiry date. They will eat it anyways. To them, it's important that you give them something. Else, they will procure the food for themselves. Yet they will still be starving. Forever.

In spite of their exorbitant appetite, they have some limits as well. Here is where you and a black hole resemble each other. Both of you can't eat everything at once.

Another thought experiment (this chapter is full of those): you are at a large dining table, maybe at an event or something. The food is exquisite, but it repeats itself so that everyone is able to reach the same dishes. What do you do? Obviously, you don't go to the other end of the table to get the exact same food that's in front of your sit – you just eat what's closest to you. Then, if you want more but already finished what was up front, you could politely ask for another portion.

Well, black holes eat what's in front of them, too. They don't get up to reach for the dishes at the other end of the table. But after they finish it, they don't ask for more. They just take. Not very polite of them, but you could keep the good stuff for yourself while providing the black hole the worst of food. It needn't even be food, really. It can be forks, napkins, plates, even the table if you don't mind eating on the floor. To them, it's the same.

That's the thing. The black hole starts with the mindset that it wants to empty the entire table. But it can't do so instantly. It begins with the closest dishes and slowly makes its way to the other side of the dinner.

Basically, they swallow what's closest to them, be it gas, stars, light or spaceships. They could potentially swallow you too. If only they were closer... But you are glad they aren't.

The closest black holes to Earth are a binary system entitled "The Unicorn", which is just over 1500 light years away. It might not seem like much, but remember that the closest star is a bit over 4 light years away. That's around 350 times bigger of a distance. In addition, the black hole is merely 3 solar masses heavy, which places it amongst the lightest of them. Its companion is a red giant with a mass similar to that.

Thus, this system doesn't pose more of a threat than any other star in the universe. Delicate black holes do exist in many corners of our galaxy, but they constitute extremely little to no danger at all.

Supermassive black holes, on the other hand, are something different. Those have gargantuan masses and dimensions, but they are much farther away from us. We are talking galactic centers here, millions of light years away. Only one of them is significantly closer than that: our own SMBH – Sagittarius A.

This 1.5 million-solar-mass supermassive black hole is merely 26 000 light years away from us, only a couple of times more remote than "The Unicorn", but way bigger. So, what about it? Could it swallow us?

Eventually, it could. But for the time being, there is plenty of stellar material at the core of the Milky Way, so it can freely feast on that. We are also spinning around our galaxy, like everything situated along its arms, and there is no sign of rapid orbital decrease (we are not really spiraling down towards the center). A subsequent suction event that would terminate our system for good might only occur in the very distant future. Galactic collisions have the potential of seriously impacting the evolution of all systems, including ours, but they, too, will only affect us after a long time. So that's not a reason to stay up at night.

Back to the origins

Regardless of their imposing size and appetite, black holes aren't something to fear of. Not during their rather extensive lifetime. But before they get to this prolonged final stage of their evolution, they generally evolve as stars – huge gaseous bodies fueled by nuclear fusion reactions within. And while this process itself sounds worrying, we know that stars don't interfere with us. They are only bright dots embellishing the skies.

However, massive stars, progenitors of stellar black holes, could be a bit more than that. You see, as they approach their end (which happens much faster than regular mass stars, by the way), the final breath differs from that of other stars.

Usually, celestial bodies like our Sun expand significantly as they reach retirement, and then expire in a great way: they undergo a nova. The outer layers of the star collapse upon it's inner ones and then expel some of their material in space, resulting in a bright object appearing in the sky, which slowly fades away in a couple of days to weeks.

Massive stars experience something similar, with the exception that the final spectacle is a grander and much more energetic one. This threshold between star life and black hole eternity is called a supernova. But these gorgeous explosions are still explosions, clearly of astronomical proportions, and there is a much darker side to them.

Just like gamma ray bursts, it's the supernovae's energy that could bring terror upon us. GRBs can also be created by supernovae, which makes them all the more dangerous, but even normal ones have an incredible destruction potential. Yet again, it's only a matter of power and distance.

Violent candidates

When compared to gamma ray bursts, supernovae account for less energy per surface area, thus being less visible from increased distances. You could think of it like this: supernovae spread their output energy over a spherical volume of space, with the resulting black hole at the center of it, while GRBs act only on small conical portions, making their energy per unit area at least 10^5 or 10^6 times greater than that of a supernova.

A supernova also lasts longer until it begins to fade – days to weeks as opposed to seconds for a GRB. This makes gamma ray bursts even more intense. Still, supernovae can be seen from extraordinary distances as well, but they need to be incredibly powerful (the most distant one was pinpointed some place 11 billion light years away). So, just like distant GRBs, they lose almost all of their punch before they get to us.

Essentially, a supernova must be relatively nearby for it to have any impact on Earth, besides the usual colorful show they put up in the sky.

The frequency of visible supernovae is pretty small as well. Part of this is due to them forming inside of the galactic arms, letting interstellar absorption seriously dim them. High resolution telescopes are able to record quite a few of them, but only the visible[11] ones could be a menace.

In the past thousand years, only 4 supernovae were observed by naked eye, suggesting a rate of around 1 visible SN every 250 years (on average). Not much, huh? And all of the ones we've seen so far were nothing more than splendid bright shells in the sky – they could also be seen in plain daytime! And they didn't even originate from far away.

One of the most likely stars that could undergo a supernova in the near future is Betelgeuse, as red supergiant in the constellation of Orion, reaching the end of its life. Its merely 500 light years away, but it's still thought to not impact us in the slightest. Funny how it could've exploded already, but we couldn't see it only after 500 years have passed...

Fortunately for us, there are no other closer supernova candidates as all of the nearby stars have insufficient mass to cause such a phenomenon. Betelgeuse is the scariest one yet – it will go supernova and will make up the closest stellar black hole to us. Interesting to think about it, but not even Betelgeuse could lead to our demise.

Conclusion

Black holes and supernovae are definitely vigorous devices in the universe's arsenal. They have incredible destructive power, far greater than anything we've ever created. Yet as of now, those world destroyers are harmless to us.

Supernovae go off all the time, really, but we are lucky enough to be in an isolated part of our galaxy, with no massive stars wishing to terminate us. We are also fortunate to be situated far out in the galactic disk, and not close to the supermassive black hole at the center, or the stellar black hole progenitors in its vicinity. That's mainly thanks to the great distances in the universe – its not only the Big Rip; expansion also works to our advantage.

So, for now, there is nothing of astronomical scales to worry about. These possibilities only made it higher up the list because they do destroy entire systems constantly. It's only that we are not one of those. In that sense, we are lucky!

But if we end up expanding our reach as a civilization, becoming interplanetary, interstellar or even intergalactic, other residences might not be as protected as Earth. But when we get there... if we get there... then what we consider to be doomsday scenarios today could be basic issues for such a futuristic species. And that's exactly what we'll consider next.

Most of the stars in the Orion Nebula (1340 light-years away) are massive and will evolve into supernovae

1. The main sequence is a rather linear portion of the Hertzsprung-Russell diagram on which are situated stars that fuse hydrogen into helium as their main source of energy. About 90% of all stars in the universe are on the

main sequence, ranging from fractions to hundreds of solar masses.

2. Velocity vs. distance diagrams, in which an unexpected trendline can be seen for all galaxies, in contradiction with observations. This suggests an abundance of dark matter as we move outwards from the center, and a great concentration of mass at the core.
3. A sphere is the shape gravity prefers, as everything tends to the same center of mass, and when particles are so tightly packed like in a black hole, gravity dictates.
4. Every object appears to have 3 dimensions, but when light from them reaches our eyes, only a 2D projection is perceived by the retina, and subsequently our brain. Basically, you know that the object in cause has 3 dimensions, but you only see them all as a 2D projection. Weird, isn't it?
5. A potential well is the region surrounding a local minimum of potential energy. As objects tend to occupy the least energetic positions in space, potential wells are their way to go.
6. Potential is the other form of energy besides kinetic. Kinetic energy is due to the motion of an object, while potential energy depends on the relative position of the system's components. In our case, we are talking about a gravitational potential energy, which is always negative – thus particles will tend towards the biggest absolute value of it, the equivalent with minimum.
7. Stable equilibrium is an equilibrium state towards the body will tend even if you displace it by a bit. The body will want to return to its original equilibrium position. Unstable equilibrium is the exact opposite – the object is in equilibrium but any displacement will make it get

further from that position.

8. The moment of inertia is a quantitative measure of the rotational inertia of a body. That's when a certain torque acts on a body and gives it a rotational velocity about an axis. This quantity is said to be conserved in every conservative system. In consequence, if a big object like a star is slowly rotating, the resulting compact black hole will spin faster – e.g. the same happens when an ice skater spins faster if he brings his hands closer to the body, making himself more compact and increasing his rotational velocity.
9. Exotic matter is a hypothetical kind of matter which is characterized by negative energy density and a negative pressure or tension. It would act like a hill in space-time from which objects will slide down, being all the more accelerated outwards.
10. The Multiverse theory suggests the existence of a multiverse, encompassing all the parallel universes in existence (of the order of 10^500). This collection of universes is only hypothetical, so the black hole-white hole gateways shouldn't be regarded as the one correct interpretation for the information issue, not until they are proved.
11. Visible as in "with the naked eye", without the use of enlarging devices such as telescopes.

I.4: The Great Filter

"Any sufficiently advanced technology is indistinguishable from magic."

Arthur C. Clarke's 3rd law

A variation of this law would be: "Any sufficiently advanced civilization is indistinguishable from God." This might upset the more religious lector, but it's not meant to disprove divinity in any way. Let's reflect on that for a second, with the first affirmation as a starting point.

What does one mean by magic? Generally, we think of magic as something inexplicable to us, something that amazes us through its surreal outcomes. We might be convinced that magicians and illusionists are merely tricking our brains into believing that what we see is broken from reality, but in the end, we are left wondering how did they do it. It couldn't have been real magic, right?

Still, magic has a special place in our minds as something we know can't be true, but at the same time we are unable to explain.

Technology is, perhaps, the key feature of any advanced civilization. And as we like to think of humankind as the most intelligent species on our planet, it's only logical to relate to technological development. So what does it this have to do with magic?

Imagine a primitive human from tens of thousands of years ago was somehow brought back to life in our modern society. What would he think of it? Most assuredly not much else than what is going on. He would be bewildered by everything surrounding him – he would have no idea.

Letting aside the fact that he just woke up from a seemingly endless coma, he would see nothing familiar to him. Everything from food, animals, houses and even humans would look astonishingly different from what he remembered. And then we get to technology.

He remembered drawing on cave walls his impression of a mammoth. Nowadays, paintings move by themselves, apparently without being influenced by any external stimuli – we call those TVs. Between the weird-looking shelters of modern humans, wind streets made out of rough material and moving boxes are roaming them. There is light outside even after the Sun has set, powered by something newly discovered like electric current. Giant metallic birds graze the skies and humanoid creatures of the same nature create more and more new stuff without ever needing a break. Nothing is the same!

What such an ancient hunter-gatherer called magic is an everyday matter in today's society. We take technology for granted, but to him it's something he couldn't even dream of. This is what a gap of a few thousands of years entails. And we are talking here about the most evolved species at that time as well. Think of what would a mammoth think of all this, or worse, a primordial cellular being!

Our technology would be regarded as magic. We would be regarded as gods! To their feeble perception of what's real, only some sort of highly superior being or divinity would be capable of performing such things. Today's humans would meet those criteria.

But humans have evolved in such a way in just a couple thousand years – a minute period to universal standards. Even more so, only in the past 200 years have we evolved significantly. For the most of human history, we did nothing but survive. Progress was made at an extremely low pace. But in the last hundreds of years we have witnessed an exponential growth in every aspect of our society.

Modern history is but a slight glimpse of the total time life had to evolve. And in this little time, we managed to achieve so many great things, that would be inconceivable to our forebears. But what if we were put in front of the same scenario as our hunter-gatherer friend? What if we were the primitive beings in the face of some other, more intelligent civilization?

Strangers in the future

We are used to being the dominant species, to having everything at our disposal. Thus, a potential situation in which we are the lesser beings is unbeknownst to us.

Think of the poor squirrels for example - they aren't the smartest creatures out there. In fact, they very often forget where they hid their nuts, so they are forced to find another tree as a hiding spot, a new burrow. Yet there are many other animals which are far less clever than squirrels.

But they do know one thing. Trees are important to them. Trees are their whole life. What could a squirrel think of us when we leisurely cut down entire forests for our own good? Our intention is not to exterminate them, but to simply gather materials, just like they do, but on a much greater scale. Even so, it's not like they can conceive this. To them, we are dangerous. We might wave and wonder at them in the park, but as a species, they see us as destroyers, no matter our intentions.

It would be truly terrible for such a thing to happen to us. Even if our homes don't get wiped out by another civilization, the sheer difference in knowledge would be unbearable. That's what we would feel if we were to be transported some thousand years into the future.

Assuming that we do not self-destruct by then, our technological development will increase ever faster, until what we deem impossible today might be a normality at some point in the far or distant future (we leave extraterrestrial civilizations for a later chapter).

Now, to be clear: technological evolution and colonizing other planets and systems is definitely the way to go. Our planet is abundant in resources and it will still take us a while before ending them, but they are limited. And for an expanding civilization that is willing to thrive for a long time to come, our general priority is to find ways to access as many resources and as much space as possible. That's why reaching out for the stars in not solely an expression, not anymore – it's a motivation.

So then, how could such a positive scenario represent a danger to humankind? In itself, it doesn't. But there are a few concepts that need to be understood before getting to our latent doom.

The Kardashev scale

In the 20th century, Russian physicist Nikolai Kardashev came up with a way to classify existing or hypothetical civilizations in the universe. His method, named after him "The Kardashev scale", is used even to this day and consists of 3 basic levels.

All of those are measured as a function of the amount of energy that can be fully benefited from at any given moment (power = energy per unit time).

Type I refers to the kind of civilization that is sufficiently evolved to be capable of harnessing the entire power of its home planet. This is something of the order of 10^16 Watts. Here we include the totality of the planet's resources. This number is similar to the amount of power constantly falling on the planet from its parent star.

It is likely that this level, the lowest one on the scale, is the hardest to attain. It first and foremost requires life to be able to form on the planet and later evolve to multicellular beings (billions of years). Those need to develop enough to have access to technology, to become intelligent (hundreds of millions of years). And then they need to evolve to the point where energy harnessing is possible at such a level (tens of thousands of years). You may think this is nothing at cosmic scales, and you would be perfectly accurate, but the existence of organic life itself is thought to be a surprisingly difficult process. After this milestone is surmounted, the next level awaits.

A type II civilization would be capable of harnessing and efficiently using the entire power of its parent star – the order of 10^26 Watts. This is 10 orders of magnitude bigger than the previous threshold, but theoretically easier to achieve in the long run.

Upon conquering the entire home planet, a civilization should aim towards spreading their seed on multiple planets; other residences with suitable conditions. Maybe those aren't possible right off, but the sufficiently advanced type I civilization is able to make them habitable via a process called "terraforming" (the term comes from the name Terra, so it might sound different for other advanced species). This way, they can live on a number of planets, becoming an interplanetary civilization. The process itself should only take a couple hundred to thousands of years, not much in comparison with intervals required for level 1. From that point on, its only a matter of resources before the species is able to construct devices competent enough to harness the energy of their stellar system.

As this step is successfully passed, another level of development is ahead of it. Type III defines a galactic civilization – one capable of harnessing the entire energy of its host galaxy. At this point, the power capacity would increase ten more orders of magnitude, getting to 10^36 Watts. This is much tougher than the precedent, or tremendously difficult to perceive, in the very least.

Even at light speed, you would only be able to travel across the galaxy in 100 000 years. We would therefore need much more futuristic engines and means of transport in order to have the slightest chance of colonizing the entire galaxy. But for an already type II civilization, such methods would be underway, and later common amongst its planetary communities. By taking one step at a time, such a civilization would be more than skillful enough to attain the rank of galactic species. Yet again, it's a matter of when.

Further levels of technological development were proposed – 10^46, 56 and 66 Watts respectively; but those are a bit out of reach for a still newbie civilization like ours. By the way, where are we situated on this scale?

Unsurprisingly, not very high up. Actually, we haven't even made it to level 1 yet. According to the formula after which the scale was proposed[1], we are somewhere around level 0.73 with our 2*10^13 Watts available. There is still room for improvement and it would take a couple hundred years of time. Our route forward would look something like this: hopeful that we don't self-destruct, we must evolve our technology – nanorobots, artificial intelligence, more efficient fuels just to name a few; at the same time we bust begin our space journey by venturing to nearby planets, asteroids and moons to examine the conditions in the eventuality of colonizing them; the most suitable candidates are the closest to us, Mars and Venus, and they could represent the first tread as a space faring civilization; while everything keeps developing here on Earth, we may eventually get to type I and be mighty enough of starting our multi-planetary colonization, by leading terraforming process where most effective; around this time, interstellar exploration might also become a reality, with manned missions to other solar systems; our expansion would accelerate and we could, in theory, roam the whole galaxy sometime in the distant future.

But for now, that's just a flight of fancy. We should focus on the little things whilst dreaming big and aiming for those essential objectives. This way, little by little, there might be a chance that we seriously climb up Kardashev's ladder.

A complementary scale not towards the very big, but the very tiny, was also created to support his ideas. That was proposed by John Barrow and details the future advancement in nanotechnology, all the way down to the very manipulation of space-time. It isn't as used as the first one, but still an interesting way of characterizing civilizations.

Another one was put forward by the famous astronomer Carl Sagan. His was not as specific as Barrow's, but rather abstract like the original scale. But instead of using power to measure the character of a civilization, he used information – bits to be more precise[2].

In Sagan's view, civilizations would be ranked from A to Z, every letter of the alphabet increasing the number of bits by an order of magnitude, with A being 10^6, whereas Z would be 10^31. We are currently around H, or 10^13 bits, though a bit below it. As you can see, there is still a long way to go, but Sagan speculated than it would be impossible for any civilization to have reached Z by now, no matter how old, as the universe itself isn't aged enough to permit information to have travelled such enormous distances so that it is available to them.

Sagan is also the one who stated this, regarding the fundamental Kardashev scale: " (humanity's behavior is) typical of a civilization about to integrate the type I Kardashev scale." Something of a technical adolescence where its actions are questionable and risky. But maybe it really is the way to go for a youngster civilization.

That's cool, but there is something about this statement that seems odd: what does he mean by "typical"?

The Fermi paradox

When used by others in the same context, this word might simply refer to the hypothetical traits of a type x civilization, hence "typical". But from the mouth of Carl Sagan, it could imply something more chilling.

Sadly, he passed away a while ago, but his works remained – exceptional as ever. And during his life, as supported by his books, he dedicated most of his time to the study of extraterrestrial life. He was a pioneer in astrobiology, also referred to as exobiology, and promoted the Search for Extra-Terrestrial Intelligence (SETI). In short, he was most curious in this particular field: alien civilizations.

So, when a word like "typical" escapes his mouth in such a serious context, we can only assume he believed in the existence of other civilizations. Maybe not candidly, but he was definitely aware of the likelihood of it. In the vastness of our universe, could it really be that we are the only advanced civilization? The only civilization of any kind? Are we really alone?

"Two possibilities exist: either we are alone in the universe

or we are not. Both are equally terrifying."

Arthur C. Clarke

You might have heard this before, and only thinking about it is exactly as it says: terrifying.

With the idea of an ever-expanding universe, in all its grandeur, we are left wondering about the chances of us being alone. Being special.

Astronomer Frank Drake came up with a probabilistic argument to estimate the number of active, communicative extraterrestrial civilizations within our Milky Way galaxy – it might not be "ours" in the end. The idea he came up with is named after him "the Drake equation", aiming to calculate the number of such civilizations as a function of multiple defining factors, allowing the existence of complex life-forms.

The first factor is the rate of star formation in our galaxy, as life would be impossible without the protection offered by stars. Then comes the fraction of those stars that have planets orbiting them. Next, the average number of planets that could support life (per star with planets)[3]. The fraction of planets where life actually develops at some point and that of planets where intelligent life is able to develop (so we can talk about civilizations) are another two important factors. We then have the fraction of those that dispose of sufficiently advanced technology to release detectable signals into space. And lastly, the length of time for which those release the signals.

These are the elements it takes into consideration. Not the most scientific of equations, but one that gives expected results – reasonable estimations. Most unfavorable results entail some tens of civilizations, while most conclusions give something of the order of tens, maybe hundreds of thousands. But anyway, that's only in our galaxy. Keep in mind that there are at least a few hundred billion of them out there!

So, in all the likelihood of us not being alone in the universe, where are the others? How come we haven't received any signals by now? This is the main proposition of the Fermi paradox.

Story has it that as famous Italian physicist Enrico Fermi bogged down in debates with his colleagues on the issue of alien life and interstellar travel on their way to lunch, one particular question of his marked everyone at the table: where is everybody?

Many more or less plausible explanations have emerged since then. Maybe we didn't see anything because of our still short span as an intelligent species, that information hasn't reached us yet. Or it could be the other way around, intelligent life hasn't yet come to be in our vicinity, so there's no way for us to detect it yet. Perhaps they are hiding from us, either they are afraid of us or are afraid to harm us, or maybe they know about us but simply don't want to interfere if there is no reason to. Or it could very well be that they are not sufficiently advanced yet.

Whatever the case, it all comes down to time. It takes time for life to evolve and reach a stage where its existence can be highlighted to other civilizations, and from that, many things can vary. And with time as background, one gruesome theory (in the true sense of the word) has arisen from all the viable versions prior to it. Today, it is known as the Great Filter.

Contextualizing the Great Filter

As the Bible suggests, at one point in the future, people will have come to be so shy of morality, that the world just won't take it anymore. And that God's wrath will bring doom upon earth, so that everything we know of will come to ruin, whilst a renewal is ahead of us...

But in the Book of Apocalypse, the word "human" only appears three times. Once in the context of something resembling human faces; once as "human measurement", so that we could get the physical idea behind the verse; and once as individuals referred to as slaves. Yet it never appears in plural form, nor does it ever denominate humans as a whole or as a species.

Because the book is deeply symbolic, especially this last part, it proved to be particularly difficult to interpret it. But in the context of us as a civilization, not much is being said. It's basically as if the Scripture doesn't necessarily imply that the apocalypse only relates to us.

Moreover, the Bible and the entire religious belief which shapes us have an overarching definition of divinity, whatever It might be. They let us believe that God is the guardian and protector of everything in this world. Why should we be special? If It is what we think It is, almighty and all powerful above everything in this world, then what would incline It towards us? The Father of the universe encompasses it in its entirety.

What I'm trying to say is that it's rightful of any potential civilization out there to have faith in It. We shouldn't be the only ones (if we are truly not alone).

With this perspective in mind, the horrendous theory of the Great Filter responds to the Fermi paradox by insinuating some sort of barrier – an obstacle that can't be surmounted by most, if not all civilizations in existence. This filter would lead them all to an inevitable or almost inevitable extinction, rendering us unable to contact them in a short enough time, as they stumble across this insuperable obstacle. This proposed fate would sound for every civilization, including our own, hence taking the paradox to a universal level: all of the millions or billions (or whatever) civilizations in the universe would live their lives with the misconception that they are alone...

You could think of this obstacle as a concrete wall on or way to supremacy. As we, or any other civilization for this matter, are trying to climb the evolutionary staircase, this concrete wall suddenly appears on one of the steps. We don't know where or why, but it does. Consequently, we are left ineffectual in front of it, for it prevents any further development of intelligent life. So, as the filter eliminates almost every civilization that encounters it, could it mean this will be our doom?

It's complicated...

We are currently standing below the first level of the Kardashev scale, and there is still a while until we get there. Essentially, we are very low on the staircase, close to the first floor. At least in theory, there is a long way to go until we are fully developed.

Still, the universe is mature enough, though not exactly adult, to allow for evolution to at least the level of interstellar spread among earlier civilizations. Maybe galactic species aren't a real possibility yet, but in the vastness of space, a decent number of type IIs should have emerged, right?

Well, it depends... It depends on the elevation of the filter – where is it situated along the staircase. Is it before our evolutionary stage (that's the case in which we are lucky), or is it somewhere ahead of us (not so great)?

Case 1: the filter is behind us

There are still many flights of stairs before we reach our maximum potential. In other words, many floors await, while we are still somewhere below ground level.

On a universal scale, the presence of a Great Filter behind us is incredibly tiny. It would, however, provide a suitable explanation to the Fermi paradox.

If the 4.6 billion-year history of Earth would be represented on a clock, all of the recorded history would be not more than a second. It took 700 million years for life to appear in its most rudimentary form – prokaryotes (aka bacterial life). Another couple hundreds of millions for those prokaryotes to learn to cooperate with one another, which led to host systems, where the inner body provided energy for the host and the host fueled the guest's mechanism, and ultimately to the most basic form of eukaryotic life (the very engine of every living organism – mitochondria). Then another 1.5 billion years before oxygen began populating the atmosphere and 800 million years more until the first animals evolved. Human beings in the form of Neanderthals appeared around 200 and 250 thousand years ago, and intelligent civilizations only 10 000 years ago. Finally, advanced society developed just some two hundred years ago, with an exponential growth in the last century. What is nowadays the dominant species on Earth is but a grain of sand in the planet's history beach.

Yet even after all this time, we have no idea how life appeared in the first place and why did it take it so long to evolve. We can only be glad it did! You see, this whole period is around half of our parent star's lifetime, and for the most part there was only rock and water, with no land, no atmosphere and primarily, no life. Sooner or later, the Sun will begin expanding, heating up our planet to a point where water will evaporate and consequently make life impossible on Earth. As Elon Musk once stated: "[...] if it had taken consciousness 10% longer to evolve it would have never evolved at all. 10% longer...".

However unlikely it may seem that there is no other form of evolved life in the universe, maybe that's the whole point. Maybe it really is an extraordinary challenge for life to even appear... Maybe even if other planets are as suitable or more suitable for sentient life, life itself isn't able to form and evolve quick enough to reach a certain level... It could truly be that we are exceptionally lucky.

We have encountered many unfortunate events throughout history, some worse than other. Five mass extinctions and many more minor cataclysms so far. Yet none of them was able to exterminate life on Earth. The 20%, 5% or whatever remained was always able to move on, continue thriving and evolve into everything we know today. None extinction event was too good of a filter – more like an imminent roadblock which we were able to overcome.

We don't know for sure what the filter might consist of, but if it weren't those catastrophic, nor life-sustaining conditions that differ from our planet's, our best bet is that some evolutionary process is, indeed, tremendously difficult to occur.

That would mean that we are one lucky civilization and have no reason to expect alien interference any time soon.

Case 2: the filter is ahead of us

In the second case, the filter is yet to be reached – maybe in a matter of years, maybe millennia, or maybe in the very distant future. Most of the proposed time periods are still like the blink of an eye in the cosmological timescale. Whatever the case, our odds are quite shabby.

A filter like none other positioned some steps ahead on the evolutionary staircase would be horrible news. Out of all the possible civilizations which went through the not so difficult (in this case) life-generating process, none have made it out alive after a certain level. But how far ahead could one such filter be?

Observations aren't on our side... In theory, civilizations could've reached a more advanced level than ours, but we haven't seen any yet. Once a specific point is attained, it is quite difficult to end a multi-planetary, perhaps even multi-stellar civilization. If it was that hard to terminate us in our most rudimentary form on one planet – cells, then how could an event be so devastating to wipe out entire planetary populations of a highly intelligent species? It's not very likely.

So, it is possible that such a filter isn't too far away from us. If none of the external dangers were strong enough to end life on Earth, conceivably only an internal disaster could be of that magnitude. The most likely scenario in this case is that there is a threshold in the evolution of any species from which there is no way back, as they are already heading towards self-destruction. It could be that in our unending quest for more resources we deteriorate our planet too fast for us to regenerate or escape it. Another depressing hypothesis is that after a civilization takes control over its home planet, it's already on the path to self-destruction. And the all present technology might be the leader of it. Whether it's a large-scale nuclear war, artificial intelligence that got out of hand or any of the future scenarios in this book, the filter can thoroughly get rid of us for good.

This is why observing extraterrestrial life might be excruciating news. If any form of life is somehow recorded in the not so distant future, in our little corner of galaxy, or even universe, then it means life isn't as hard after all. That would make us as insignificant as we thought at first, not special in any way, and above all, on the edge of a knife.

Prokaryotic or eukaryotic life-forms alone wouldn't be great. Complex animal life would be worse. And anything similar or lesser than us would be terrible. It would mean we are close to the end. If we see alien ruins we can just start crying...

In this case, the best chances are if we see a number of more advanced civilizations, interstellar or galactic ones, which would imply that if there is a filter, it is still way ahead of us. But then again, we haven't seen any in spite of their theoretically more advanced technology. And it would be hard for such a civilization to self-destruct...

It could be that the best we can hope for is that we are alone in the universe, however arrogant that would sound. Because if the filter is ahead, you can be sure it's not going to be friendly with us.

The no filter scenario

What if neither case is correct? What if... there is no filter after all?

The third and last possible scenario is that maybe no filter is at the middle of the Fermi paradox. We might be alone, or we might not be, but either way a Great Filter is not the reason for our lack of evidence regarding alien life. It could very well be that signals from them haven't reached us yet and neither us, nor them, are evolved enough to communicate efficiently in our young universe.

There is no limit for our technological development and overall evolution as a civilization. And there isn't one for others either. It may be that we are destined to live our lives, self-destruct or continue advancing on the evolutionary ladder, and get to be a joint galactic or even extragalactic society, prospering alongside other species from other parts of the cosmos.

That would be nice, wouldn't it? And this should be our major goal for the future. We must work together and share a strong motivation so that our civilization can at least integrate into a galactic community, if not create one itself. Not a single thought regarding a filter of such nature should stop us from aiming for the stars! It might not even be real to be honest...

And even if it is, it is most probably behind us or very close ahead. The former would be wonderful. Even the latter should inspire us, and not prevent further technological development. As I've previously stated, if a barrier is waiting for us, it will most probably be caused as a product of internal errors. For this reason, we must learn to cooperate and work in the advantage of one another, and not to undermine the others in a desperate attempt to prove that we are better. It's the only way we can pass a potential filter.

So, assuming that we will eventually learn to do this, not as a restricted community, but as a world-wide civilization, there is no stopping us! Or is it? One final proposed theory is a Great Barrier – similar to a Great Filter, only we know exactly where it's situated.

This concept appears in the context of continuous expansion. As we develop more and more with the purpose of leading simpler and better lives, our general motivation and progress are supposed to exponentially increase (or that's what they are supposed to do). So, if we manage to colonize other planets, other stellar systems and even other galaxies, at some point there will be nothing left to do. The technology will be at its absolute capacity, the universe fully occupied by us or some other form of life we might evolve into and everything that we could ever think of will be possible. It would be as if the genie from the lamp gave us infinite wishes. You would use 100, let's say 1000 of them if you let your imagination run wild, but what do you do next? What could you possibly wish for other than what you have, which is everything?

"What happens when an unstoppable force meets an immovable object?"

popular paradoxical question

In our situation, the unstoppable force would be the immense, god-like human potential of an extremely advanced civilization, and the immovable object would be the very universe which limits us. One could argue with the application of this paradox on multiple grounds.

First, after achieving such a potential, what stops us from going even further and explore all the universes in existence. Basically, everything that God created should, at least in theory, be explorable. We would leave no stone unturned. Everything that was in the realm of mystery, uncharted, would then be just a touristic objective for the future inhabitants of the world. But as the word "theory" implies, all of that is merely a hypothesis.

Even if that were the case, what would we do next? After the entire multiverse is explored? Maybe there's something beyond that? Yes, maybe. But everything should have a limit. Our universe has limitations, the hypothesized multiverse too. It is counter-intuitive to go on forever. And if it truly went on forever, then everything would repeat. Over and over and over again. It would be quite a boring future. An eternity in which we will just live our pitiful, yet perfect lives, without changing anything.

Second, it may sound unbelievable for our current, very bounded selves to think of a future civilization – our successors – as attaining god-like potential. By definition, God is above all there is, and no matter how advanced we are, It would still be capable of infinitely more than us. So, in truth, there must be a limit to evolution even from a spiritual perspective.

We will never be able to reach the level of divinity, no matter how hard we try and no matter how much time elapses. And we won't even get close to it. There will always be a barrier than simply can't be crossed.

So, in the end, the fate of any infinitely advanced civilization is unequivocally determined. There will be a point where nothing else can be done. We will just carry on with our boring, pointless lives until the universe (or whatever our world will be by then) comes to an end. We cannot be gods. And we cannot persist forever. You may want to reflect on that for a while...

Conclusion

The subject of an imminent limit to humankind in particular is a rather philosophical one. But from my point of view, such topics are where the communion between science and belief can be most evidently underlined.

Frankly, there is no telling whether any sort of filter is a viable theory, which can be applied to the real world, or just another hypothesis made to excuse our inability to detect extraterrestrial life. But regardless of the existence of such a filter, the issue proposed towards the end of this chapter is real, and we can all relate to it in one way or another.

However, there is still a long way to go, as repeatedly emphasized. We have to deal with our little earthly problems first: learn to cooperate as soon as possible, try not to self-destruct and work our way up the evolutionary scale, little by little.

We should stay optimistic and aim for the best. And above all, be glad that we've made it so far, yet not be too comforted by that as to be lazy.

For all of those reasons, the probability of a Great Filter is clearly an existent one, a miserable one as well, but uncategorical. Thus, we can only hope it isn't definitive and deemed to extinguish the human flame, while I categorize it as the fourth on the likelihood scale of doomsday scenarios.

"The future world will have to find

equilibrium in the technology pendulum swing."

PSALMS OF HORROR

Stephane Nappo

1. The formula used to calculate the Kardashev scale index is the following:

K=(log P – 6)/10, where P is the power available, and K the civilization level.

1. A bit is the most basic unit of information in computing and digital communications. It comes from "binary digit", 1 or 0, representing the logical state. 8 bits are called a byte, which is more generally used when measuring information (on your phone or any mobile device, which has a number of Megabytes – Mb or Gigabytes – Gb).
2. Planets found in the so-called Goldilocks zone of the star system – the region where water is possible in liquid form, as water is considered to be the foundation and primary resource for organic life.

I.5: Earthly abnormalities

"Forces of nature act in a mysterious manner. We can but solve the mystery by deducing the unknown result from the known results of similar events."

Mahatma Gandhi

This is all science is based on. We must wonder, we must doubt and be curious enough to uncover the hidden nature of reality. That's the beauty of it. As famous biologist Rupert Sheldrake put it: "Science at its best is an open-minded method of inquiry, not a belief system." And this applies to everything there is.

In order to get the gist of what's going on, we have to interpret the results from the magnifying glass of previous events. The idea that only experience leads to learning is, in fact, very true. Not once in history have we discovered something new just by thinking that might be – we always experimented, tried new things and made connections with other experiences. All of that, which distinguishes us from most machines, are the fundaments of human learning – trial and error, and not just a given set of information which is put into practice in a perpetual manner.

And where better to experiment than our own planet?! Yes, this infinitesimal drop of water in the giant sea of the cosmos. But this little blue marble (as referred to after the well-known Apollo 17 picture was taken) has much more to it than we could imagine. And before beginning our space exploration, we should keep in mind that there are many uncharted territories even on our tiny marble.

Yet as tiny as they might be, currently some of them pose much greater threats than anything the outer space can hit us with. And from now on, most of the doomsday scenarios left to cover will focus on mundane issues, either nature-related or of anthropic origins. Each and every one of them could have catastrophic implications, but what we'll call Earthly abnormalities throughout this chapter are still not the most concrete and concerning. And due to the lack of scientific consensus on specific problems that made it to this category, they will all be counted as highly unlikely scenarios for an eventual doomsday event.

Earth's Kevlar vest

First and foremost, we must understand that our planet doesn't wish for our extinction just as much as the universe does. It is, however, quite a hostile environment, with aggressive attacks being landed upon us all the time. But concomitantly, there would be no life without it. We needed its atmosphere, its water, its vegetation – that was the only way we could make it this far.

So, one might argue that we are stuck in a love-hate relationship with our home planet, and we can't escape that (at least for now). And this works both sides: sometimes we love it, sometimes we act as if it were our archenemy and intentionally or not merge our forces to harm it the most; Earth, in turn, constantly nourishes us, but also comes up with though challenges like it wanted to kill us.

Now some may say that a love-hate relationship is the worst you could be involved in, others that it is the best. When it comes to Earth, I think it's something in between. We don't mind because we live, but aren't too fond of it either. And one of the essential features of this ongoing relationship is the Earth's magnetic field.

As of our prior discussion, the main element of our Solar System, the Sun, not only gives us light needed for suitable temperatures, photosynthesis and other processes, but it also keeps planets in orbit and above that, it protects them all from hazardous cosmic radiation via its magnetic field. And while it does extend billions of kilometers, it isn't impenetrable and it really isn't that strong (many stars have magnetic fields way stronger[1]).

So even with the Sun's protective sheath, noxious energetic particles still make their way to us. And if that weren't enough, the Sun itself sends numerous other harmful particles in all directions. With all of that radiation directed towards us non-stop, how come we are not dead?

Well, that's thanks to Earth's protective envelope, which is a similar magnetic field to our Sun's. I like to think of it the following way: a soldier inside a tank will still wear a Kevlar armor, even though the tank offers him plenty of protection. Similarly, Earth still has its own "Kevlar vest", despite the Sun's shielding. So how does it work?

Just like a bullet-proof vest, it's the inside that gets the job done. For the Sun, differential movement between poles and equatorial regions generates the magnetic field. For the Earth though, which isn't "that gaseous" as the Sun, it's the changes in fluid portion in the Earth's outer iron core which are thought to generate the field[2].

This way, invisible magnetic field lines exist and reenter Earth after encircling it in outer space. The behavior of the lines is exactly that of a simple magnet's lines. You know, those red and blue magnets which have a north and south pole – Earth has them too. But you still can't see the magnetic lines. What you can do to visualize them is to sprinkle iron filings around the magnet, and they will arrange in the form of the magnetic fields inside the plane. And that happens in any plane: semicircular magnetic lines existing the north pole and then entering the south pole, all the way round the magnet.

Those field lines encase the entire planet in an invisible blanket, in the shape of an ant's body. Now, don't freak out, I know it sounds weird, but bear with me. The front of its "body" is in the form of a flattened sphere, more like an ellipsoid. This is the part of the field headed towards the Sun. The backend is quite elongated, the field lines being stretched out the farther they go from Earth. This looks similar to the back side of an ant – a bit sharpened like a needle.

This distortion from a uniform, toroidal field[3] happens because of Solar wind. As the energetic particles come from the Sun, racing towards us, they deform the magnetic field imposed by Earth, but it still resists. The Sun's magnetic field is also stretched due to the same reasons: as the system moves in the galactic disk, cosmic radiation coming towards it changes the shape of the heliosphere (spherical, hence the name) into an oval, thus extending less in the moving direction.

Back to Earth. So, the magnetic field is rather elongated, but that doesn't mean that energetic particles are able to get to the planet's surface. Actually, they are driven away by this field, just like air currents are driven away by an aerodynamic vehicle. In our case, Earth is the vehicle. They follow the field lines and aren't able to get too close to us.

This way, they are detained above our planet. One could say that this magnetic field of ours acts like a trap – "magnetic trap" that is. As a consequence, certain portions of the field which capture more radiation were called Van Allen radiation belts, after the physicist who discovered them. Two radiation belts were found, between 1000 and 12,000 km above ground level, and from 15,000 to

60,000 km respectively. The existence of a third belt mainly composed by electrons like the outer one is debated. Also, the ranges of the other two belts aren't too precise, so results vary. But one thing is for sure: they are giant prisons of inmate energetic particles; therefore, satellites demand additional protection in order to operate at those distances (otherwise the radiation would fry their circuits, rendering them useless).

Another consequence of the magnetic field is the formation of boreal and austral aurorae. That's because field lines are more concentrated around the magnetic poles, which are somewhat close to geographic poles. And as the high-energy particles get to those areas, luminous light spectacles can be witnessed in the vicinity.

Not bad so far. Then what's the whole deal with Earth's magnetic field?

Poles shifting?!

There is nothing wrong with our magnetic field. It is quite small, yes, but it does the trick – it manages to protect us from harmful radiation. Au contraire, it's what is happening with the field that we could find worrisome.

We saw how GRBs, supernovae or other cosmic hazards can alter the field, but something else, more mundane is altering it as we speak (well, as you're reading in fact). It's the very movement of the poles themselves which weakens the field.

To be clear, magnetic poles, unlike geographic ones, are constantly moving. It's just that nowadays, they were found to be moving more than ten times faster than usual. Just like a magnet moving about its axis changes the position of the iron filings, the Earth's magnetic field lines will also change. Only the Earth is a gigantic magnet and the effects could be way worse. And this rapid, accelerating movement of the poles suggests an early attempt of polar reversal.

What that means is that the poles get interchanged, such as the north becomes the south and vice-versa. The reversal itself wouldn't have too much of an effect on us, other than maybe our compass needle going crazy and animals depending on the magnetic field like pigeons or other birds becoming disoriented. The process, on the other hand, would have an immense potential of negatively affecting us.

The last polar reversal took place some 780,000 years ago. It's only strange how that polar reversal happened twice as long ago than the usual interval between such events. On average, two reversals would succeed in a matter of 300,000 years. With that in mind, the extended time since the last reversal, as well as the increased movement of the magnetic poles, the next event of this nature might happen pretty soon.

The aftermath of a polar reversal

During anterior polar reversal events, the magnetic field of Earth went haywire. It was as if "the magnet" transformed into multiple magnets. And contrary to the expectations, this didn't strengthen the magnetic field, it actually decreased it. Basically, any time a reversal happens, the magnetic field decreases to about 10% its normal value during the process.

And it's not like the process takes a day or two. It may take years, even centuries. And this entire period will be characterized by an extremely weakened magnetic field. During this time, Earth and its neighborhood will be significantly more exposed to dangerous radiation from the Sun.

First off, satellites will begin to feel the weakening as radiation will perforate their shielding. Their circuits may fry and errors will arise, leading to them ultimately stopping from functioning and falling down towards Earth. Then, animals may get disoriented – migrate in weird directions and suffer collisions.

And last but not least (and definitely worst), every electric circuit will break down, irreversibly. Streetlights will turn dark and no more light pollution will interfere. Nights will be as dark as they were before the invention of the light bulb. TVs, phones, computers and the other electronic devices will fail. No more phone calls, no more limitless information at your fingertips, no more nothing. Ovens, toasters, mixers, refrigerators, light installations, water filtration systems, heaters all of those will suffer the same fate. Cars, buses, planes and other means of transportation will also break down, and bicycles and trains might become the fastest ones again.

In a nutshell, everything we got so used to in our lives will no longer be available for at least some hundred years. For most of us it is challenging to even imagine such a time, and I assure you that the reality might be much worse than that. The entire humankind would be dragged back a few centuries, something like the dark ages. Maybe we won't return to our primitive lifestyle, but something along the lines of the 17th-18th century would be a viable scenario.

It might not kill us all, that's true, but the damage would be massive and irreparable for at least the time of the reversal. Besides the direct victims of this change, radiation could cause mutations in the human genome, leading to way more cases of cancer and other diseases. The best way to go would be overprotective sun cream and no more sunbathing; additional clothing items and umbrellas would help as well.

The economic impact would also be huge. Today, if the power went down in the US, the Government would need to spend over 40 billion dollars a day. Trillions more dollars for a short blockade of the Suez Canal or the impact of hurricanes and you get the idea of how much 100 or so years of electronic stone age would cause.

So, while it may not be the exact reason for an apocalypse, a disaster of this scale and duration would certainly damage humans like nothing before. Add the failure of all electrical circuits to intense mutations, economic damage and random riots bursting out and you get a decent recipe for catastrophe. Good luck with that!

"They were not allowed to kill them but only to torture them for five months. And the agony they suffered was like that of the sting of a scorpion when it strikes. / During those days people will seek death but will not find it; they will long to die, but death will elude them."

Revelation 9:5,6

Internal movements

The good part is that the Earth will return to its original state immediately after this process is finished. Humans will finally be able to overcome this dark era and move on – develop even further towards the next stages of evolution.

But what if a magnetic field shifting phenomenon occurred not due to the constant motion of poles, but because of the effective generation mechanism? In other words, what if Earth's core weren't up to such a task anymore?

For as long as we can estimate, the planet's core was always fit enough for sustaining differential liquid motion inside of it and generating a decent magnetic field around the rock surrounding it. So, while the field fluctuated from time to time, it was always present nonetheless.

But as part of the Earthly abnormalities chapter, I am going to detail the slight chance of the field disappearing completely. And the major trigger in this case would be the core halting its internal movements; cooling down and essentially stopping.

It's important to understand that, at least to our knowledge, such a thing has never happened. The core was always hot, with a temperature like that of the Sun's surface. Yet it's a known fact that it is constantly losing energy, by releasing heat either into the mantle[4], or to the surface of the planet. Other systems behave similarly, but for this one in particular, it's not an actual possibility to receive heat in order to prolong its activity. All the bonus energy is that gained from Earth's rotation and tidal forces.

Studies conducted by a team of researchers examined the evolution of a heat-conducting metal, called bridgmanite, most commonly found at the boundary between the core and the mantle. But their study yielded strange results. As it appeared to them, the Earth seemed to be cooling much faster than they initially thought – a trend observed in other rocky planets as well. They believe this is mainly driven by the intensified dissipation of energy in the mantle, leading to a more radical decrease in the core's temperature. However, a study isn't enough to tell for sure whether the core's cooling accelerated, but it does highlight the risk of an eventual halt happening faster than anticipated.

An activity of such proportions will likely not happen in the near future, so we'll be safe from that for years to come, millions of years really. Who knows where humanity will be by then... But for the sake of argument, let's look into how a phenomenon like this might impact today's society.

A sudden halt

It will definitely take a while until the core stops spinning, so the effects would appear gradually; but on astronomic time scales, it would happen quite suddenly.

A lower magnetic field will be the first thing you'll sense. Again, birds losing control, misfunctioning compasses, weirdly positioned aurorae and perhaps error in satellite measurements such as GPS, weather, TV and so on. Then, all your electronic devices and home appliances will go crazy (not all at once, but rather little by little). Increased rates of cancer and mutations will be obvious in the long run as well.

Nothing special by now. It's looks like a loyal copy of the previous scenario, where the poles reversed. But one essential difference is the much more increased duration of this developmental drawback. The other differences are even more significant.

A fundamental difference from the other scenario is that in this one, the magnetic field won't just decrease by some amount from its original value, it will disappear for good. That means no protection against cosmic radiation and solar winds whatsoever – more energetic particles will reach the surface. Now, you might think: but don't we also have an extra protective layer, the atmosphere? We do, but not for long.

In the absence of a magnetic field, solar winds will not be damped by anything. They will be strong enough to blow everything away, including the entire terrestrial atmosphere. That means no protection, no greenhouse effects for retaining solar energy and, above all, no oxygen. We would suffocate to death, freeze to death or die from cancer – whichever happens first.

Because the core cooled down, volcanic eruptions won't be an issue anymore, as there is not enough energy to pump up the magma. Also, hot springs won't be hot any longer and basically everything on Earth will cool down. Our planet will become a frozen wasteland just like Mars.

Mars, too, has no atmosphere, no liquid water and no magnetic field, although it once had all of those. As its core turned solid, it is possible that this exact scenario once happened to our red neighbor.

As no life will be possible above ground, our bets alternative is to live below it. Entire ancient cities were carved deep below the surface and could host tens of thousands of people each. Why couldn't we do the same?

We could. The problem is that we need to find a way to survive in such conditions. Ancient civilizations often went outside for resources, in spite of their Hermitian lifestyle. We will not have that luxury, as there will be nothing out there.

So, with no life and almost no reason to scavenge for resources, humans will be forced to live their tough lives underground, with plenty of oxygen, food and water supplies gathered beforehand. At least we'll be pretty much protected from outside's hazards...

But even so, we won't last for too long with such limited provisions. Both the surface and the underground will soon be devoid of life, unless we find a way to either preserve the magnetic field or make our way to other planets, more suitable for life.

Millions of years from the instauration of nothingness, it is possible that Earth's rotation will engage the core again, stimulating the internal movements. Consequently, the magnetic field will return to its former glory.

What about the atmosphere? The inner blanket would take even longer to regenerate. After the core restarts its activity, things will begin to come back to normal. That includes volcanoes and hydrothermal vents, which will then release carbon dioxide in the soon-to-be new atmosphere. At the same time, the Sun will begin to decompose water vapor into hydrogen and oxygen, which will also make it in the atmosphere. Long after the renewal of the magnetic field (and even longer after we are gone), the atmosphere will slowly but surely form again, however in much thinner layers.

An unbearable atmosphere

So, let's wish that such a thing doesn't occur to soon (although it probably won't) and that our atmosphere will stay rich and healthy as we know it. But be careful what you wish for. Too much of a good thing could potentially lead to negative consequences. One of those unfortunate scenarios will be presented in what's to come.

Our atmosphere is extremely helpful to us, besides acting like a cosmic defender. It also provides humans, and basically all organisms, with vital resources. And by far the most important out of them is none other than oxygen, the catalyzer of aerobic life.

Currently, the atmosphere consists of about 21% oxygen, 78% nitrogen and a few other elements. This is just the right amount – not too little so that we can't breathe and not too much so that it would kill us. But things haven't always been like this... Now get ready for a story!

Some 3.8 billion years ago, life in the most basic form, that of bacteria, first appeared on Earth. Every microcellular organism, if we could even call them that, was thriving in the planet's oceans, without too much of an interest to evolve. Back then, iron was plentiful in the oceans (almost the entire surface was covered in oceans), which is why their color wasn't the nice, deep blue we are used to seeing today, but rather green[5]. That was because of the lack of oxygen in the atmosphere. There was a lot of nitrogen, water vapor, carbon dioxide and methane, but no oxygen whatsoever. That worked just fine for primordial microorganisms who didn't rely on oxygen to survive – they used other chemical elements such as sulfur or carbon dioxide instead.

Eventually, though, this happened: the first aerobic lifeforms began populating Earth's oceans. Obviously, nothing too complex, only what's known as cyanobacteria. Anyway, those new microorganisms related on the famous process called photosynthesis. It was essentially a way in which they could procure their own energy as long as exposed to light – solar radiation. Through this process, these bacteria would also release some amounts of oxygen, just like today's plants do, just not as much. However, the rapid creation of these led to huge quantities of this element being thrown out into the oceans and atmosphere.

The oceans turned from green to a blood red, while anaerobic organisms began dying out because of oxygen intoxication. Then, as it spread into the atmosphere, it reacted with primordial greenhouse gases, drastically changing the Earth's chemistry. A swift global cooling followed, the result being a 300 million-years long ice age, known as "snowball Earth".

In spite of all those, life went on and continued its gradual evolution into reptiles, then mammals, primates and finally, you, me, your neighbor and every other person. So, it was an unpleasant event to say the least, causing a mass extinction that wiped out 99% of all organisms... but it led to us and every other being on this planet, so we shouldn't complain. But what if something similar happened to us, as in "today"?

Becoming sick of oxygen

One thing is certain: an increased amount of oxygen today wouldn't kill nearly all life on Earth as it did back then. Not directly, anyway. That's not to say that a hypothetical scenario in which oxygen somehow increased to a whole other level wouldn't be bad for us.

As biology advocates, a higher percentage of oxygen in the Earth's atmosphere would give rise to a wide range of effects on all living organisms, including humans. And those effects are situated at opposite sides of the spectrum – some very good, some very bad. We'll have a look into those.

Insects are the ones that would benefit the most. As opposed to mammals, insects breathe through some tubes called trachea. More oxygen wouldn't cause any damage to them. On the contrary, it's expected that this would help them improve their organism. According to previous experience and also the study of their bodies, the more oxygen they breathe, the bigger their size. Of course, they wouldn't grow indefinitely, but they would definitely be much larger than what we are used to. In previous eras, when the oxygen made up only slightly more of the atmosphere than it does today, spiders and various other arachnoids were the size of birds, while dragonflies were the size of falcons or eagles. Who knows how much bigger they could grow? It wouldn't be fun for those suffering from arachnophobia.

But for the faint of heart, there is some consolation. Humans would also gain new abilities. We would overall be an improved version of ourselves. As almost all of our energy comes from oxygen that gets to brain and muscles, the rest being procured from food and water. So, with abundant oxygen, our senses would sharpen and our muscles would be in their best shape ever. More strength, more speed, more endurance, too. You would also become more resistant to colds and other outside dangers – that would be thanks to increased oxygen quantities in your blood vessels.

Then, there's also the dark side of things. For humans and most animals, increased levels of oxygen might lead to serious health issues. Everything from DNA alteration to oxygen intoxication (what caused so much death a few billions of years ago). You would have superhumans strengths, but they wouldn't make up for all the damage that's caused.

Oxygen is also the stimulant of burning reactions. You know candles burn for as long as there is oxygen to consume, which is why if you place an upside-down glass over it, at some point the flame will go out. The same thing is applicable to mostly everything. In the presence of so much oxygen, all crops would catch fire, wood would burn down and also forests and practically all vegetation would burst into flames. And yet again, bodies of water like oceans and seas would suffer, just like in the past – they would begin oxidizing. So, life on Earth would be nothing like we know it...

The reverse of the medal

"The horses and riders I saw in my vision looked like this: Their breastplates were fiery red, dark blue, and yellow as sulfur. The heads of the horses resembled the heads of lions, and out of their mouths came fire, smoke and sulfur. / A third of mankind was killed by the three plagues of fire, smoke and sulfur that came out of their mouths."

Revelation 9:17,18

The biblical prophecy of Revelation's chapter 9 is strikingly resembling the above scenario – a world filled with fire, smoke and death. Even the sulfur is present in the fragment. And according to that, a great part of mankind could be ended by the merciless fire. This goes hand in hand with the scientific background of this process. But for your peace of mind, something like a "great oxidation"[6] isn't meant to happen for the foreseeable future.

There is still another oxygen-involving alternative: another only slight variation from the fine-tuned conditions that allow life on Earth. Surprisingly or not, this one refers to a lack of oxygen in the atmosphere, which could also prove catastrophic to us.

Oceanic anoxic events describe periods when large portions of Earth's oceans were depleted of dissolved oxygen. This implies toxic waters for both us and marine species and has the potential to destroy most marine life and severely impact our development on all planes.

The main cause of these types of events is the fertilization of water bodies, stimulating the growth of phytoplankton. As those die at some point, microbes decompose them, using oxygen in the process; oxygen that was vital to marine lifeforms. The unfortunate ones, those living in areas with low oxygen are more sensitive to disease, have trouble growing, can't reproduce and even die because of it.

In the past, anoxic conditions were linked to extreme volcanic activity, resulting in a build-up of carbon dioxide in the atmosphere, which led to increased temperatures, causing an accelerated introduction of nutrients due to water's cycle in nature, which in turn led to the uprising of phytoplanktonic life. Those led to numerous reductions in the number of marine species and even a mass extinction some hundreds of millions of years ago, when the majority of species died out.

So, what stops such an event from happening today? Well, nothing really. It's only up to us whether this will happen or not, but by the looks of things, we aren't doing anything to prevent it. Anoxic events are strongly connected to the slowing of ocean circulation, climatic warming, and elevated levels of greenhouse gases, according to most geologists and oceanologists. Does any of that sound familiar? Of course it does! We are contributing to it on a daily basis.

As scientists Denise Breitburg said: "nutrient inputs from human activity – things like nitrogen and phosphorus – coming from fertilizers, farmland, sewage, fossil fuel combustion", all of those contribute to the deoxygenation of oceans. They also led to global warming and increased temperatures have been shown to stratify the water column, so that oxygen from the atmosphere is no longer able to enter the cycle.

As this happens, marine species situated in coastal areas are heavily damages, but those in open water aren't having a much better health either. Even if they try to escape, more often than not, fishermen will "take care" of them. Drinkable water resources might also be damaged by this sort of activity, so everything will come down on us in the end.

And this process is increasingly more dangerous as it's becoming wide-spread, which was exactly the case in the last decades. This problem might accentuate to the point where damage is irreparable, some time in the near future. That's a bit worrying.

But we should remember that the only way we can prevent this from happening is by being more cautious with our everyday activity. Not so much pollution, not so much waste and not so hasty development and we should be fine. The future is in our hands, and we sure don't want to cause our own doom quicker than needed.

Conclusion

Our planet is mysterious, to say the least. There are so many unknowns in Earth's equations, so many nooks and crannies left to be discovered. And some of them are surprisingly overwhelming even for the most intelligent life form, humans. Such were the possible and somewhat probable apocalyptic scenarios presented in this last chapter of the first section.

No matter what we do now, there is still no way to prevent those disasters (except for the last one). And there is little warning that they are coming. Luckily, at least those specific doomsday events progress gradually. We would have some time to prepare beforehand, but they would still damage us considerably.

One thing might sooth us, though. It's unlikely for most of those to happen in the very near future (again, the exception being the last one). Some are threatening us that they could trigger at any point, but most are not even possible at the moment. The thing is that if we have at least some hundred years to evolve, each and every one of them will be much easier to overcome. Solutions to the magnetic field problem will appear with time. So will appear for controlling the atmospheric consistency. It's just a matter of time and will. Because if we want to develop, we will develop, and what seems like an end of days today, might become a basic issue for tomorrow's society.

In conclusion, if we are lucky enough to avoid these scenarios in the near future, they won't pose any threat at all in what's to come. They might, after all, happen at any time, but it is unlikely that they will fully exterminate us, and quite hard to believe that humankind will not come up with some sort of method to attenuate their impact.

For this reason, Earthly abnormalities made it to the top of the lowest section. We should hope they won't occur too soon, but at the same time (and that's important) take care of our planet while preparing cooperatively for the worst.

"I cannot do all the good that the world needs.

But the world needs all the good that I can do."

Jana Stanfield

Earth's aurorae – events caused by the magnetic field

Forest fires would be much more common in an overly oxygenated atmosphere, with even wet leaves and wood on a cloudy day catching fire relatively quickly

Petra – a city carved into rock (a type of shelter in case of a serious magnetic field disruption event)

1. Dwarf stars, neutron stars and especially magnetars have magnetic fields much stronger that our Sun's.
2. This happens in accordance with something called "the dynamo effect". In short, the liquid iron in the core is moved through an already existing, weak magnetic field. This produces a current, which generates another magnetic field which interacts with the core material, giving rise to a secondary field. The two fields than add up to form an even stronger, general magnetic field. The mechanism is able to keep this field working for long time periods.
3. A torus is a 3D shape generated by revolving a circle about an axis in the plane of the circle. This way, you get a shape like that of a donut, with a hole in the middle, or, in our case, a degenerate hole as the axis is basically tangent to the circle so to say (it's easier to imagine it that way). So, the magnetic field does have a ditch in the middle on either side (close to the poles), but there is no actual hole.
4. The internal structure of Earth is as follows: inner core, outer core, mantle and crust. The mantle and crust may be imparted in multiple ways, depending on different layers. But we are interested in the fact that the mantle is what comes right after the outer core, and is the portion in which most of the energy would be dissipated in the analyzed case.
5. When oxygen was lacking, green rust was all over the place, giving oceans a color of green; later on, when oxygen was abundant, rust was rather reddish, hence the blood red color of our planet.
6. That's the term used to refer to that period a couple billion years ago when oxygen led to a mass extinction.

SECTION II:

Unpredictable but unlikely scenarios

About more possible events that could end us all if taken to another level

II.1: Earthquakes and tsunamis

I was sitting quietly in bed with some of my friends, playing card games late at night. A quiet night. The last of a three-days-long journey. We were at a Physics competition, if I recall correctly. And after so much exam-taking and anticipative waiting for the results that were about to be announced the next morning, the only thing our brains were able to cope with was a relaxing card game. A better alternative might have been sleep, but we really wanted to persist through the night and spend time with our fellow colleagues from around the country.

Besides us triggering the smoke alarm with bug spray in the middle of the night (twice!), it was, as I've said, pretty much silence around our not so friendly-looking accommodation building. Everyone was exhausted and ready to give up and go to bed, but me and my team were fresh as always – or at least that was the impression we wanted to give.

Then at some point, around 3 AM, it was my turn to shuffle the cards. So, while struggling to stay awake, I gave them the best shuffle I could. I began dealing the cards and by the time I was done, a weak tremor encompassed the room. Cards began quivering. So did the beds and so did we. We were a bit confused at first, then I calmly stated: "Hey guys, there's an earthquake going on." No one paid attention, as they were way too tired to focus on both the game and my words at once. We resumed our game, but then one of us, with a 15 second delay, kept telling some other to stop shaking the bed. He quickly excused himself, but no one believed him. Then I restated my point, claiming that it's an earthquake; still, everyone else kept blaming that guy for shaking the bed, even though we were spread out across three beds and two separate tables – there was no way he was shaking all of them.

As tired as we were, we waited for the waver to pass and moved on with our game. It was only the next morning when we were more surprised to find out that it was, indeed, an earthquake than our contest results. Fortunately, it was only a slight one, with a 3.5 felt magnitude at 100 km from were it was generated, that was where we were standing. We were quite lucky, but those closer to the central point suffered much more damage.

To us, it was an interesting experience, we've never felt a significant earthquake before. But this kind of experience isn't that uncommon, to some people's surprise. In fact, if you were born on planet Earth (and presumably you were since this book made it to you), you've most certainly experienced at least one earthquake in your life. Maybe not a strong one, but you did.

For the extraterrestrial readers, the same thing could be deduced provided that you come from a rocky (telluric) planet. If one discovered time travel and went way back to the early years of Earth, earthquakes would be interfering with him on a daily basis – and strong ones for that matter.

But I am not here to talk about alien life, not in this chapter at least, and time travel is still a far cry for humanity, assuming it is even possible. Instead, I wish to talk about these modern days calamities which are earthquakes, as well as their implications.

Before diving into this complicated subject, however, I would like to explain myself as to why I dedicated an entire chapter to this discussion.

Out of all the apocalyptic scenarios you've heard of, this one is probably not the most commercialized, and definitely not the center point of too many prophecies. However, the striking difference between earthquakes (and tsunamis) and the other theories elaborated so far is just that – the others are just theories.

We don't know what would happen if the universe simply fades out of existence, as we'll probably be long gone before it happens. There is no telling exactly what a GRB snipe would lead to. And truly, we still have tons of blank spaces regarding black holes, let alone what would happen if one were to devour us some day. Even with these so-called "earthly abnormalities", which we know for a fact that they might and will happen (especially the polar reversal scenario), humans have never been through such an event. Ever. Polar reversals and their implications, for instance, were disastrous to whatever life was there on Earth. This has been going on since the dawn of history. But in the last short hundred-thousand-year period when we can talk about humans for the first time, no event of such kind came about. Let alone recorded history, or even modern history (when the outcomes could be really terrible).

For earthquakes it's a whole other story. They are going on constantly and have been doing so for as long as we can recall. Even more so in the early days of our planet, but let's get back to our times. The point is that unlike for the other scenarios, we are able to see the direct consequences of this one.

Everyone hears something about them in the news, whenever an earthquake of bigger proportions hits. We can see with our own eyes the damage induced by such a natural disaster: buildings collapsing, the ground shattering and people taking shelter in desperate attempts to stay safe. It is by no means pleasing to the eye, but it's the harsh reality. Earthquakes of various degrees go off on a regular basis (we'll see how regular a bit later on) and, to our knowledge, there is nothing sophisticated enough to prevent them from happening. This is one of the reasons why it made it to the second section of this book.

The second reason is regarding a biblical standpoint. You see, while it isn't the center of attention when a doomsday warning spreads across the internet, things take a complete turn when analyzing the Bible. It might be because of the extended background of this calamities. It might as well be due to catastrophic evidence of earthquakes in the old pages of human history. One thing is certain though: it is by far the most tackled possibility which could(!?) bring upon the end of days.

You might notice the unconventional punctuation between parentheses. This sign, "!?", is known as an interrobang, and it is generally used to express both awe and doubt towards a certain statement. This is exactly the case with the last phrase of the previous paragraph, particularly with the word "could". It's both the great lengths to which the Bible goes in order to express the potential devastation of earthquakes, and also the predominant question of this work – could it really happen?

"I watched as he opened the sixth seal. There was a great earthquake. The sun turned black like sackcloth made of goat hair, the whole moon turned blood red,[..] and every mountain and island was removed from its place."

Revelation 6:12-14

"At that very hour there was a severe earthquake and a tenth of the city collapsed. Seven thousand people were killed in the earthquake, and the survivors were terrified and gave glory to the God of heaven."

Revelation 11:13

"Then there came flashes of lightning, rumblings, peals of thunder and a severe earthquake. No earthquake like it has ever occurred since mankind has been on earth, so tremendous was the quake."

Revelation 16:18

Those are merely a few of the denotative references to these extreme natural phenomena in the Book of Revelation. Other symbols and metaphors are widely used throughout the Scripture to describe earthquakes at their most pernicious.

But even in the Bible, the scales of these cataclysms aren't as impressive. It doesn't take an infamous earthquake for a tenth of a city to collapse, especially in those early times when the Book was written. 7000 isn't a death toll that great compared with the hundreds of thousands of lost lives in modern deadly earthquakes. And last but not least, no earthquake in the history of mankind was able to deal significant damage to an entire continent even. It would take one of extraordinary measures to endanger all life on Earth, just like the Bible implies.

So then, the question stays: how big need an earthquake be to exterminate our species? And is there any chance that one such thing could occur in the future? These are only a couple of the questions we'll answer in this chapter's analysis. So, brace yourself – it's going to be shaky!

A monster in the depths

Seconds before I began shuffling the cards, deep down in the heart of the abyss, the monster makes its presence felt. And it's not as if he woke up and started crawling towards us, but rather he snored.

It was only for this slight disturbance in its otherwise calm sleep that triggered something I personally never dealt with before: an earthquake.

But let's pause for a second and rewind to the origin of this "snore". The actual mechanism is quite intricate and we'll return to that in a bit. Before that, we shall focus on how the monster got to us so quickly.

As we later found out in the morning that the earthquake was produced some hundred kilometers away, we were left wondering how come it was powerful enough to shake the beds and shuffle the cards once more. Well, at least I was. My older colleagues who had already studied a bit of seismology in physics class were able to clear my mind, explaining why the quake hasn't lost too much of its intensity by the point it got to us. And now that I've meticulously studied the issue myself, I believe I could shed some light on it.

First of all – why the shake? Why did the beds move in the first place? It all comes down to the nature of the "snore". We all know that a snore is nothing else than a sound, maybe annoying at times, but it is a sound. A sound that makes it to one's ears, causing a chain reaction which leads to reception by the brain; the brain then gives the sensation of hearing, so that you can perceive the snore. But how does it get to your ear, because this is the trigger of that chain reaction?

Every sound is fundamentally a vibration in a medium. Any medium, whether it's gaseous, like air, liquid, like water, or solid, like metal. Provided that it vibrates, it can induce that vibration to your eardrums and later excite the receptive cells inside the ears and create the sensation of sound. But at its base level, sound is just an oscillation of that medium's molecules. We thus say that the medium vibrates. These vibrations are regarded as sound waves, as they propagate from one point to the other, away from the sound's origin, determining the direction of the sound.

But sound waves are not the only form of waves. There are many others, but we are mainly interested in mechanical waves, in which category sound waves are also included. However, they can't really be seen. You can hear voices in the air, as vocal cords induce an oscillation in this medium. You can also hear sounds while underwater or when putting your ear on metallic surfaces, but you don't actually see this kind of mechanical waves.

Now as an experiment, how about you take a bottle of water. Tilt it slightly and you might notice a weak motion until water settles at its new equilibrium position. Tilt it faster and the motion intensifies. Shake the bottle and the motion becomes chaotic. It's as if water wouldn't settle. That's because you induce this wave motion in the liquid. Consequently, it begins to oscillate depending on how you proceed with the bottle movements – this is a wave you can see.

Similarly, many parts of our environment benefit of the aspect of mechanical waves. The giant rock we live on is no exception. Being a regular solid medium by nature, of course it allows for the existence of mechanical waves. These propagate, like all other waves, and get to us. That's why the beds moved back when we were playing cards, because the ground below them oscillated. Only, there's a twist.

What do I mean by that? Seismic waves manifest in a distinct manner to what you see when shaking a body of liquid. Here we get to our second point: earthquakes are a collection of multiple oscillations in the medium. All of them originate from the same spot, called the focus, similar to a sound-producing source, but spread in different ways and at different velocities.

P waves come first. These are often referred to as "primary waves" or "pressure waves", hence the acronym. What they do is they compress and expand the medium in the direction of propagation, instigating this back and forth ground movement. This provides an explanation for why the beds quivered in just one direction at the beginning.

Then, the quiver intensified as S waves (or "secondary/shear waves") came in. These ones were of greater amplitude, in spite of their lower velocity. While the waves themselves traverse the same as P waves, the particles in the medium vibrate up and down, perpendicular to the direction of propagation.

Both of these oscillations were of little amplitude and could also engage air particles and subsequently produce sound, there is another type of waves left to cover, one that only travels underground, but which poses the greatest threat. We are talking about the "surface waves" – those that have much greater amplitudes and typically arise when the source of the earthquake is close to the planet's surface. Often described as the aftershock due to their significantly lower velocities, surface waves can behave in two ways: either as "ground roll", ripples like those of water which cause bodies to move up and down (this effect can be seen over larger spatial expanses), or as chaotic horizontal waves, shaking objects randomly and more thoroughly in directions perpendicular to wave propagation.

This latter form of ground vibrations was seen about 30 seconds after the original shock, when the beds began shivering from left to right as well.

Stubborn puzzle pieces

"He who strikes terror in others is himself continually in fear."

Claudius Claudianus

This monster hiding deep down below the land of man is trapped in a seemingly endless conflict which causes him to "burst out" from time to time. Even though he is sleeping, he is actually restless. He is struggling to get a good sleep, but occasionally, its belly wishes otherwise. Perhaps this seems a bit odd in a serious context like that of earthquakes, but you will soon see that the analogy is pretty accurate.

As peaceful as planet Earth might look like from cosmos, we know for a fact that's not the case close-up. What not all of us know is that below this unnerved layer, called the Earth's crust, our planet is gone rogue, worse than what could ever be achieved above ground.

If you cut an apple in half, it may strike you that its peel is extremely thin compared to the pulp. You could think of that peel as the crust. Below it is a vast volume of molten metals, acting anarchically as temperatures approach that of the Sun. Each layer of this giant apple's pulp plays its own role, but one specific outer layer is responsible for wreaking havoc on Earth for billions of years, and it's not likely that it will stop soon: the asthenosphere.

Hotter rock closer to the center moves upwards towards the surface, replacing the cooler material at the top (as cooler material is heavier). Therefore, in this molten rock fluid of the asthenosphere, convection currents[1] are created. These occur all over the place and the movements look like ovals from a section[2].

The convection currents push the crust as they roam the asthenosphere, which is why a rumbling, agitated stomach would be an appropriate depiction for the ongoing activities of our monster. If only it were that simple...

So, if Earth's surface, practically floating on the asthenosphere, were merely a spherical sheath surrounding the molten interior, that the motion incited by the currents would be nothing but a shy back and forth swing of this thin layer as a whole. Here comes the tricky part. The crust is, unfortunately, not that smooth foil wrapped around our planet.

It is rather like a former picture of a perfect planet, which ended up cleaved into multiple components (as nothing is perfect in nature). Under the given circumstances, one might say that Earth's outer layer is a gargantuan puzzle, in the form of a sphere[3]. The puzzle pieces are, in fact, huge, rigid slabs of rock, called tectonic plates, left loose on the asthenosphere. On the bright side of things, they fit almost perfectly and only tolerate modest displacements – a few centimeters per year. This is a very small number compared to the size of Earth, but all of them have an unquenchable tendency to move around. In a sense, they are just like us when desperately trying to find that ideal sleep position. It's just that they don't. None of them do.

Henceforth, in their enthusiastic urge to not stay put, tectonic plates give birth to the most feared of the biblical catastrophes. By colliding with one another, sliding sideways or moving away from another, tectonic activity leads to earthquakes of varying proportions, apart from other side effects of each specific type.

Most common earthquakes are produced as a consequence of sliding events. There are tens of these every day, summing up to around 10 to 20 thousand a year. Usually, the smaller plates suffer from this, and not the major continental plates[4]. Because of that, these earthquakes aren't as powerful as the other types.

A second way earthquakes may occur is if two plates tend to get separated from one another. These are quite a bit more intense, but they are not to blame for the most destruction. Long-term upshots can also be noticed for these tectonic events, mainly divisions of large pieces of land. Many went on in the past in the process of continental drift, but we can experience them nowadays too. The biggest known case is the Great Rift Valley, in the eastern African region. A few millions of years from now, the horn of Africa will rupture from the main land body, and the newly created space between them will slowly be flooded by the Indian Ocean.

We've saved what's worst for last. Here come the subduction zones, the most substantial of fault lines. These come about when two plates, usually major ones, are both moving in opposite directions, causing head to head collisions. In the end, one conquers the other and comes up on top. Deep sea trenches or high mountain ranges stem from these events, but so do horrific earthquakes. There's no need to say how powerful these are and, truth be told, it's not known how great they can get. But there does seem to be an upper limit to earthquakes.

Climbing up the Richter scale

Curiosity is what sparks human development. And for as far back as we can trace the history of modern man, people have been seeking answers to both mundane and unfathomable questions. Some were looking up at the stars, while others were more ground bound, but they were all investigating natural phenomena in the search for innovation.

As previously mentioned, the Holy Book of Christianity accounts for earthquakes as being some of the primary concerns of early society. It was only reasonable that people struggled to find prevention or, at least, warning methods for when an earthquake was about to strike.

Some evidence suggests serious attempts even from Antiquity, but we are more than sure of primal technologies for detecting earthquakes in the last two millennia. At first, they were based on the little stress caused by seismic waves from far away, which could help people detect where an earthquake was coming from and estimate the intensity of it; though that wasn't of too much use as the earthquake would have already happened by the time the not-so-sensitive detection systems were able to record anything significant.

With time, detection methods perfected themselves. Seismographs and computer-based systems devised a brief prediction method for incoming quakes, but we'll cover forecasting possibilities in what's to come. It's the invention of the seismograph itself which helped scientists come up with a means for earthquake classification.

Before that, all estimates were based on the outcomes of these events, nothing concrete regarding their actual power. No sooner than 60 years after the invention of the seismograph had seismologists formulated a comprehensive scale for the magnitude of earthquakes – the well-known Richter scale. Unlike its precursors, this scale was based on the physical properties of seismic waves, taking the logarithm of the amplitude as definitory for the level on the scale. Thereby, the scale grows exponentially, each magnitude increment of 1 corresponding to a tenfold increase in amplitude[5].

A rule of thumb applies here: the sturdier the quake, the less likely it is to occur. Basically, more than half of the earthquakes are of magnitude 1-2 and even lower. These cannot be felt naturally by humans and go off all the time near micro-tectonic fault lines. Magnitude 3 earthquakes aren't uncommon either, but they do trigger less often. Magnitude 4 and above occur at local hotspots, like the one I've experienced back then. Level 5 is where things get real – weaker buildings take damage from these earthquakes and a significant shake can be felt. At level 6 most structures are harmed by the quake, whereas at magnitude 7 everything gets out of hand. These cause great destruction over a larger area and can already be detected all around the globe (as waves propagate). Magnitude 8 are catastrophic from every point of view, but happen only once or twice per year, only around the most serious of fault lines (where continental plates meet). The last level of the scale is 9, at which point destruction is unimaginable. The strongest earthquake ever recorded happened in 1960, near the coastline of Chile, at the prime subduction zone on the planet. Its magnitude was estimated at 9.5-9.6 on the scale...

Technically, the Richter scale may go on even more, but no direct evidence of stronger earthquakes is available. However, there's lot of speculations with regard to Earth's early activity, back when it was only a proto-planet. Some 4 billion years ago, as the first signs of tectonic plate structure showed up, earthquakes were more abundant, and presumably more impactful as well.

So, the early structure of our planet was permissive to earthquakes, but as oceans and continents began forming, it wasn't as prone anymore. Tectonic movements as we know them, and as they have been for a decent while (billions of years, that is), shouldn't allow for too extreme quakes. A fair and intuitive restraint for modern earthquakes would be somewhere around 10 on the Richter scale. But what if I told you that not a 10, not an 11, not even a 12, but a 13-magnitude earthquake occurred just a couple million years ago?!

I would have to admit that I am a bit cheating here, but hear me out. 65 million years ago, an external stimulus brought about the death of 75% of all life on Earth. That was when the dinosaurs became extinct. The widely accepted cause of this tragedy is a 10-kilometer wide asteroid, which struck what is currently the Yucatan peninsula, leaving a 180 km crater behind. Yet again, this might not seem like much, but the energy released was hundreds of thousands of times bigger than the most destructive earthquakes ever recorded (the equivalent of a magnitude 13 earthquake). This utter energy was released in the forms of mechanical waves of all sorts, including seismic waves felt worldwide in the form of a gargantuan earthquake.

But the dinosaurs didn't die just because of this shock, there were many other factors which contributed to their peril. So, we are stuck with this question: could a single earthquake, even of this size, destroy an entire civilization like ours? Maybe not on its own...

Flooded by disaster

The energy released in the collision with the dinosaur-killer asteroid was, indeed, the equivalent of a magnitude 13 earthquake. However, it is important to understand that this enormous amount of energy was later dissipated in multiple ways, only one of which is the peculiar earthquake we are interested in. Because of this, the magnitude most certainly didn't reach 13.0 on the Richter scale, but rather somewhere around 11 or so – not as much as initially believed, but still outperforming any other that we are aware of.

In a manner of speaking, not even cosmic projectiles the size of cities are able to quake us to death. It would require an even more colossal impact to cause a planetary tectonic disaster. On this basis, the manuscripts throughout the whole Bible are not exaggerating on this issue, not at all. It is totally feasible for an earthquake to destroy entire cities, even with modern infrastructure. Same goes for magnitudes: it isn't out of the ordinary for a major one to occur once in a while. So, even if they are the foreground of biblical calamities, they are not portrayed as most devastating neither by the Bible, nor by science.

But earthquakes are not the sole ingredient of the recipe for disaster...

The movie which initiated the series of events which then led to the writing of this book is an exquisite demonstration of how disasters may escalate. What first seemed to be a merciless earthquake capable of demolishing entire cities, maybe stronger than anything before, but nothing to fear of at a planetary level, later manifested through volcanic eruptions and tsunamis of unprecedented dimensions.

All of these phenomena are tectonic related, but the former deserve a spot in a higher category, thanks to their greater destructive potential. On the other hand, tsunamis aren't shy either. They could cause wider devastation and claim many more lives than usual volcanic eruptions, but strictly from a doomsday perspective, tsunamis aren't as likely to be the cause for it.

As the movie also put it, earthquakes are usually the catalyzers of these oceanic predators. It is because of underwater seismic waves that energetic bodies of water come to be. Many of these earthquakes are unnoticeable, but some – those forming in high activity fault lines (where continental and oceanic plates meet) – are strong enough to exert their seismic energy in the form of underwater tremors and tsunamis.

In a nutshell, these tsunamis are a series of waves caused by displacements of significant volumes in a large body of water. They propagate like any other form of mechanical wave, in all directions. An exception can be noticed when they get close to the shore (a body of land acting as a barrier in its propagation): the wave speed decreases, but the amplitude reaches higher values. This is characterized by an abnormal height of the waves, reaching tens and even hundreds of meters skywards.

Like I said, earthquakes, and tectonic activity in general, are infamous for producing some of the tallest tsunamis ever encountered. But, more often than not, they are accompanied by other factors such as landslides, the calving of glaciers or even asteroid impacts, as previously analyzed. These can further contribute to the amplitude increase of the already disproportionate waves, leading to even more intimidating sizes.

Extreme examples spin around values of hundreds of meters tall. An Alaskan tsunami caused by a 7.8 magnitude earthquake, in tandem with 90 million tons of rock dislodged into the bay, topped 520 meters back in 1958. That was the largest of its kind to ever be recorded. It was like a moving wall the size of a skyscraper, closing in on you from all directions... a terrible sight for sure. Yet bigger doesn't always mean worse.

This skyscraper-tall tsunami was responsible for the lives of only five people, whereas the notorious 2004 Indian Ocean tsunami took the lives of 230.000 and only measured 30 meters as it headed inland. That one was caused by the third-largest earthquake, with a magnitude of 9.2-9.3 on the Richter scale. What was special about this one was that the energy released was tens of times stronger that the Alaskan tsunami. Additionally, it occurred in a densely populated area, so that it destroyed entire cities over 14 different countries, and had the longest duration of faulting ever observed, between eight and ten minutes. The conclusion we draw is that the energy is paramount in our discussion.

Atlantis – more than a myth?

In the works of the famous Athenian philosopher Plato, Atlantis was a fictional island mentioned as an allegory to the major social conflicts of the time. Legend has it that this great city ended up submerged into the Atlantic Ocean upon falling out of favor with the deities.

Everyone has heard of this mythical island, continent even, swallowed by the waters, and it continues to be trending. On the same note, many have contemplated on the existence of such an underwater city. But as no concluding evidence has been brought to light, the common result is a disregard for this possibility. Surprising as it may sound, I am about to change your mind!

Well, not really. I am not aiming to persuade you into believing that Atlantis is real after all. Rather, I wish to emphasize the science behind it and put forward a few explanations as to why we should think twice.

An unsettling phenomenon that has been accelerating in the last decades is the rise of sea levels all around the planet. The primary source bringing about such increases is global warming – worldwide climatic changes powered by anthropic activities. These have many implications among which is the melting of glaciers, directly rising sea levels.

While major changes can clearly be seen in the structure and span of glaciers, it didn't seem that sea levels pose that much of a threat. Until now. It is estimated that, at the current rate, sea levels will increase 2 meters by 2100. It may be hard to estimate the consequences, but think of it like this: every seaside area of land is at high risk of being flooded. Huge cities like Kolkata or Shanghai and even large portions of countries such as Denmark, Bangladesh, the Netherlands or Indonesia are some of the main targets of rising water levels. This could endanger hundreds of millions and their only chance is to fly away from their slowly submerging homes.

"So God said to Noah, "I am going to put an end to all people, for the earth is filled with violence because of them. I am surely going to destroy both them and the earth.""

Genesis 6:13

In antithesis[6] with the Book of Revelation, the Book of Genesis portrays another doomsday event, not encountered again throughout John's Apocalypse. It paints a great flood, at the beginning of human history (though not enough information is given to pinpoint it precisely), in which all life known to man was eradicated, as a result to increasing violence amongst individuals.

There was, however, one righteous farmer, by the name of Noah, who has received divine advice and proceeded to build an ark to save the species of Earth. Many were drowned, but Noah and his family, as well as pairs of land mammals, arachnoids and birds, were redeemed from the cruel fate and helped perpetuate the flame of life.

No similar encountered was witnessed anywhere else in the Bible. By the looks of things, this was the worst flooding event in human history. As a later excerpt reiterated, God stated that "To me this is like the days of Noah, when I swore that the waters of Noah would never again cover the earth." (Isaiah 54:9). But if this is the case, how come we have close to zero evidence on the topic of the Great Flood?

At this point, a parallel discussion about Noah's Ark would be welcome.

From a physical perspective, nothing mentioned in the Bible regarding this Ark is implausible. Structurally, its characteristics would present the perfect balance between comfort, strength and stability, which would make sense provided it was designed with the guidance of God. According to the biblical dimensions, the Ark was supposed to measure about 170 meters in length, 28 meters in width and 16 meters in height. It was also said to be the greatest timber structure ever built, enormous by today's standards as well. In truth, it was compared with giant modern ships, like the Titanic, although not quite. But keep in mind that modern ships are no longer built with timber, they are made of steel, and they also include lots of utilities which were futile for a purely survival boat, like Noah's Ark. He probably didn't have a pool or a built-in mini-golf course on his ark.

In fact, a comprehensive simulation run by Korean scientists proved that an ark corresponding to the biblical description would be more than suitable to roam the oceans.

Another quandary might arise: how did all those animals suit on the boat? You may be tempted to think that millions of species must have been crammed inside, but at no point does the Genesis mention anything of all species being fit inside. What it does say is that two animals of each "kind" were suited on the Ark. That would rather refer to a more generic family of animals, and not the species themselves. In that sense, dogs, as well as wolves and even foxes were considered of the same kind, so that only one pair needed be accommodated. The number would drastically reduce to a couple thousands in the worst-case scenario, with birds and arachnoids occupying little space anyway. Also, evolution has come a long way since then, so who knows what species used to thrive at the time.

So, once the physical debate is settled, we are still left with the lack of evidence. But actually, evidence supporting the existence of a Great Flood and an Ark of this sort is available. The most prominent discovery is an ark-like structure, identified beneath snow and volcanic debris on Turkey's Mount Ararat. Geological records seem to agree with the biblical timeline, and the shape and dimensions of the discovered frame are consistent with the description of the Ark. The dots are starting to connect. Furthermore, the position of the archaeological discovery coincides with the geographical area where civilization was concentrated thousands of years ago.

Here is when the fun part comes in! Intense studies carried out over the last decades demonstrated the existence of prehistoric settlements under the Black Sea and nearby regions. A particular ancient site discovered near the coast of Bulgaria helped scientists determine how sea levels changed in previous millennia. It was thus established that a sudden increase in water elevation occurred some 5000 years ago, roughly the same time as the Bible implies for the Great Flood.

In the end, it might really be possible that the Greek legendary city of Atlantis was real, under one form or another, and that the Great Flood was truly a calamity of immense proportions. With all the evidence, there is but one dilemma which still stands out: how could something flood the entire planet?

The Bible points out pretty clearly that all species were to be eradicated at the time of the Great Flood: "I will wipe from the face of the earth every living creature I have made." It wouldn't be that difficult to wipe out beings from certain regions such as the Black Sea, but when amplified to a global level, things take a wild turn.

While it is believed that the flooding of the Black Sea (actually that increase we were talking about) was of tectonic origins, it most probably came from something else rather than a tsunami. The Bible suggest forty days of continuous rain, but that could be a simple consequence of increased sea levels and not necessarily the direct cause for the flood. This is because new water expanses will increase the humidity of the atmosphere with their evaporation, causing eternal rains to vanquish the rest.

But for the entire Earth to be swallowed by waters, something more than rain should happen. Sea levels do increase rapidly, though not as fast to worry about all the continents submerging.

There is one clue from the past which is particularly helpful in our debate. If we look into the geological timeline of our planet, as uncertain as it may be for certain points in the past, it becomes obvious that continents weren't always the way they are today – continental drift caused countless changes throughout Earth's eras. And sometimes, pieces of land not belonging to any of today's continents (i.e. which are underwater now) played a major role in the existing terrestrial formations. Alternatively, what we are familiar with today might have belonged to the realm of waters millions of years ago. There were specific periods when almost no land was risen above sea level. This may strike you as odd, but it wasn't because of heavy rains or global warming. All of those activities were of tectonic nature.

This means that, unless we find innovative solutions to this problem, we will be facing the same issue sooner or later (over millions of years, but still). So, if nothing intervenes by then, world-wide water levels could change in report with land masses, such that almost no land will be visible anymore. Everything will be underwater. But this surely won't be spontaneous like in the Bible.

So, back to our everlasting question. If neither for the rain, nor the sea levels, nor the continental drift, then what could cause a flood of apocalyptic proportions? We are only left with tsunamis...

The fury of waters

Whenever a tsunami is about to hit, a trustworthy indicator is the retreat of the sea-line on the coast. This is due to the ground settling at the site of origin, while waves slow down and gather up to form the incoming water wall. The deeper the ocean, the taller the crest of the tsunami. At the same time, the farther water recedes, the stronger the tsunami.

In case you see this happening at the beach, make sure you move as far away as possible from the shore, or seek high spots on taller buildings to protect yourself from the fury of the seas. It will be a matter of minutes, but still more than the time available in case you hear a warning of a tsunami about to strike.

Assuming you get to safety in due time, make sure you stay away from the shore for a while. Numerous times have people not listened to this advice and ventured back surfing and sunbathing, only to find out that a second smaller-tsunami came about.

The idea is that tsunamis, like all other waves, have a crest-trough configuration, the major difference being the so-called wavelength (the distance between consecutive crests or troughs). For tsunamis, it is hundreds to thousands of times greater than usual waves, ranging from 100 to several hundred kilometers. Because of this, whenever the first, accentuated crest hits, be sure that a second or even a third are coming. Depending on the magnitude of the tsunami, your best bet is to stay safe for a few hours following the initial strike.

The point this short survival lesson makes in our discussion is that should an enormous tsunami hit today, it could be days to weeks before everything comes back to normal. Of course, with a tsunami is wouldn't be possible for the whole Earth to be covered with water for too long, as the general water level doesn't significantly increase and the tectonic changes can't get that major. A few weeks would probably be acceptable though.

In "2012", tsunami waves reached heights of 7-8 kilometers, engulfing significant portions of the Himalayan range. For waves as monstrous as that, 4 weeks, as the movie put it, would be a decent timeframe for things to settle (however, water didn't really recede at that point so we are left wondering).

In that case, it was all due to sudden tectonic movements due to an accelerated Earth's core. But as we've pretty much ruled out that possibility for the formation of waves this size, just one scenario remains. That is, whether something big enough could displace a volume of water in such a way that kilometric tsunamis are generated. After all, the world's biggest tsunami on record was created that way. But that time, the mundane conditions were favorable, and it still only measured half a kilometer or so. The mightiest candidates for this grandiose job are once again asteroids.

65 million years ago, when the dinosaurs met their harsh fate, a 10 km wide asteroid impacted Yucatan peninsula. As the water was generally shallow in the area, it could be considered that the asteroid hit land, and all the fallout that happened afterwards was due to that. Still, a portion of it made it to water, causing 100-meter-high waves all around.

A common thought is what might have change if the asteroid were to hit deep water, somewhere in the middle of the ocean, or even the Mariana Trench (the deepest on the planet, with comparable dimensions to those of the asteroid). The direct destruction might have had less implications in the short term, but in the long run, giant tsunamis would've encircled the planet for days or weeks. Estimates go as far as 5-kilometer-tall waves due to an oceanic impact of that kind. Now we're talking Great Flood!

Not as high as the Himalayas, but enough to produce a flooding incident with no precedent (maybe except for the Great Flood itself). It is challenging to make any claims regarding the intensity or duration of such a catastrophe. What is certain is that this is perhaps the only way to stir a Great Flood-like apocalypse.

Conclusion

You might have been alarmed by the topic of this chapter, but keep in mind that we've only debated the most excruciating scenarios regarding earthquakes and tsunamis – everything was taken to the extreme. In reality, it is extraordinarily unlikely that any such event could scale up to a universal level. The worst seismic activities ever recorded didn't even come near, and severely influenced only small portions of the globe.

The main idea behind classifying them above the other scenarios presented so far is the continuous confrontation between us and Earth's rage. Even so, the frequency at which they occur isn't as high as one would imagine.

Earthquakes do happen a few times a day, but the wide majority are so weak that humans can't even perceive them. Tsunamis, however, are even less frequent. The rate at which they happen is about four per year, and only one in ten years raises to above 10 meters. Furthermore, these only come to be in specific areas. You will never see them in the middle of a continent, nor close to smaller bodies of water like seas. Almost all of them take place in the Pacific Ocean, in a region called "the ring of fire", where major tectonic plates meet.

One drawback from our perspective – the victims – is the unpredictability of these phenomena. If it wasn't for this, they wouldn't have been considered threats to our species. But we've devised some methods to give us an advantageous position.

Besides the usual receding of oceanic water prior to a tsunami (which doesn't happen all the time), there are a few other ways we can better analyze the moment one is meant to strike.

The name itself translates to "harbor wave" from Japanese, just because of the late warning signs. There were stories of fisherman standing in the direction of the wave who felt nothing, but when they returned on the shore, everything was wiped out by a tsunami. This is due to the rapid increase in the waves amplitude as it advances inland, as opposed to the tiny ripples farther out in the ocean. For thousands of years has this been a problem, but lately, new detection systems emerged, able to tell when an earthquake is coming based on these ripples they create far away from shore. The most accurate information about the threat is provided by tsunami warning centers, where they record the swifter seismic waves, minutes or hours in advance, depending on the origin of the tsunami. This way, by the time the slower moving tsunami waves hit, people would've already fled the place.

For earthquakes, it's a bit different. For the most part, scientists count on long-term forecasting: when and where did earthquakes happen in the past. It might not sound promising, but quite useful information can be drawn from this, as to predict the time and location of eventual intense seismic activity. Sensors are often used as well, but they don't reward us with too much time for preparation. Apart from that, the most telling sings of a quake might be invisible to these sensors.

New interpretations of first-hand results revealed a particular abundance of radioactive elements (radon and thoron) before a tectonic process of this sort is about to go off[7]. On that basis, complex detection systems involving these emissions have been proposed, especially in high seismic activity areas.

For the time being, though, we are to rely on our imperfect warning systems, which still provide us with better chances than finding ourselves trapped in the middle of a natural disaster without any previous caution whatsoever.

"The poetry of Earth is never dead."

John Keats

Destruction caused by a magnitude 7-8 earthquake

The outcome of a medium-sized tsunami (close to shore)

Artistic illustration of a kilometric tsunami

1. Convection currents form because a heated fluid (Earth's molten rock in the mantle) expands, becoming less dense. The less-dense heated fluid rises away from the heat source, while pulling cooler, heavier fluid down to replace it. The fluid at the top then cools as that at the bottom is heated, giving rise to this cyclical motion which drives our planet's tectonic movements and much more.
2. Because of the cyclic nature of the process, the molten rock seems to be following a closed, round trajectory in the asthenosphere, thus looking something like ovals or cylinders.
3. Earth is not actually round, but rather a flattened sphere because of its rotation about its axis. This gives rise to a slightly bigger radius at the equator than at the poles. A more accurate description of its shape would be an ellipsoid or, better, a geoid.
4. There are seven major plates: African, Antarctic, Eurasian, Indo-Australian, North American, Pacific and South American. These give rise to the most active fault lines, close to the coast of Japan, the western coast of North and South America, as well as in the Indian Ocean and in the Mediterranean. Then there are medium, small and micro-plates, which are still able to generate earthquakes, though usually not that big.
5. The logarithmic scale implies a tenfold increase between successive magnitudes. As the scale revolves around exponents of 10, whenever this exponent increases by 1, it follows that the intensity increases by 10 (a magnitude 5 quake shakes 10 times as much a magnitude 4).
6. By "antithesis", I invoke the fact that one is the first and the other is the last book of the Bible – one is about creation

(but not only that as you'll soon see), while the other is about destruction, about the end of humankind as we know it.

7. Small fissures in the Earth's crust as it prepares to give birth to a quake allow these elements to escape and make their presence felt in the atmosphere.

II.2: Alien invasion

I fondly remember my first interactions with the field of astronomy. I used to watch lots of documentaries and struggle to get my hands on any space-related book my eyes came across. I would stay up at night to movies and videos on the internet and then double-check the facts in books and other sources I found enjoyable as a kid.

Everything shedding light on cosmic issues was fascinating to me at the time; and still is, but back then I would get mind-blown several times a day. And it's for those daily experiences that I began my own "reflecting sessions" as I call them. The name might suggest a much deeper meaning to it, but from an external perspective they wouldn't seem as inspiring – it was just me, laying in bed or, conversely, going in circles around the room, and mentally amplifying what I've just learnt (usually in the dark). For me, however, it was truly enlightening and relaxing at the same time. Those were the only moments when I frankly felt a connection with the transcendent, as I was thinking about whatever came to mind in the grand scheme of things. I've picked up that practice and kept employing it as a recourse ever since.

Nowadays, I most commonly use it when finding myself in difficult or stressful situations, but the metaphysical side of it is often present as well. So, even though my usage of reflecting sessions could be regarded as abusive, it hasn't strayed from the straight and narrow. The most prominent proof of this claim is the everlasting debate between my optimistic and pessimistic selves on one particular, widely controversial topic.

Assuming that you are unaware of the chapter title at the top of the page, I will continue my deliberation with the grounds behind the prolonged nesting of this subject in my mind.

At many points in life have I publicly claimed that the wonders of cosmos had sparked my interest in astronomy. And while that is partly true, it still doesn't justify my lack of closure in statements like this. Poetic as it sounds, these "wonders of cosmos" normally lead one towards planets, stars and galaxies. Someone presenting interest in astronomy may go as far as black holes or nebulae, but not much else. All of those are, indeed, beauties of the universe. It's just that they merely describe old-fashioned specks of it, to put it that way.

In truth, it is not the cover of this giant book of cosmos which triggered my hunger for astronomy. It's what hides behind it. And just like in the case of a book, the content is much thicker than the cover – there's far more to be discovered than what we currently know.

So, the part I usually leave out for some reason, is that I am most attracted of the unknown, the mysteries which lurk out there. And one of those hidden variables is none other than the existence of extraterrestrial life, which makes up the point of this ongoing debate inside my head.

Honestly, no wonder it does. Every children's book, cartoon show and video game had to involve aliens one way or another. From those teaching you about celestial bodies, to the Lilliputians that would show up on Mars or on the Moon, aliens were pretty much omnipresent in a child's perception about space.

Then came the movies and the fiction books, where the portrait of extraterrestrials significantly darkened. No longer were they seen as friendly green creatures. It was the exact opposite – aliens became macabre, anger-driven beings, acting to their own interest, which mostly coincided with our worst nightmares. One began to see the other side of things and their eagerness to get in touch with aliens severely dropped with age.

But these feelings of horror typically had their apex sometime in adolescence, maybe a bit further on. Adults have lost a great part, if not all of their interest in the topic, recalling past memories only once in a while when some eerie piece of news makes it to the spotlights. Yet when they do, they tend to ponder. Do aliens really exist? If yes, then how come we haven't received any sign from them? If not, then how do you explain all those outrageous discoveries on the news? But then again, if they do, would they intend any harm? And so, even the most stable adults have a slight chance of spiraling out of control as they take a closer look upon the subject.

It could be too harsh of an expression: people going crazy about aliens is not that common of an encounter. But if there's one thing to remember from this book is that reality is harsh. Merciless even. On the other end of the spectrum, however, is a personal favorite: "the devil is not as black as he is painted". It suits this chapter best, accurately foreshadowing the upcoming parley, an it is applicable for the entirety of our doomsday scenarios. So, instead of breaking down from the possibility of an alien invasion, remember that.

"You can't hide behind your finger."

Old proverb

Not knowing...

By definition, aliens are extraterrestrial beings. Just as the name implies, these creatures represent something apart from our planet, Terra. In the common belief, they refer to humanoid creatures, meaning they would present similar traits and behave in a close manner to ours. But this is the popular perception taken to the extreme. The norm is either tall or short, slim, green creatures, with big heads and dark, elongated eyes. Tell me if I'm mistaken.

Whether you think of them this way, or another, you and I and anyone ever depicting aliens are most probably wrong. This is completely understandable. It's clear that our reasoning may be flawed from a variety of factors. But the one, most obvious factor is our absence of evidence.

You might remember our discussion from chapter 4 of the previous section, when I justified our inability to predict the existence of a Great Filter by emphasizing the young age of intelligent life on Earth. At the same time, even if we were here for longer and had more time to evolve, there would be no better means of estimating an eventual filter than direct observation of how it acted on other intelligent species. This is where we fall short.

Until now, no concrete arguments were brought towards the presence of extraterrestrial life elsewhere than our planet. Not even in its most basic forms like bacteria, let alone sentient life. So, because we are neither technologically advanced enough to reach out to "others", nor lucky enough for "others" to signal us, everything involving aliens is subject for debate (and may continue to be so long after this book is released). This is why I classified alien-related doomsday events as unlikely, but also unpredictable.

Therefore, I am not expecting too serious consideration of any supposed proof of alien encounters, nor of any too precise description of extraterrestrial characteristics and behavior. But a purely objective analysis from the prism of our limited knowledge and evidence wouldn't hurt anybody.

Speculation is all over the place. Some interpretations are more prevalent than others, yet the norm stays the same: slim, big-headed, dark-eyed creatures. From time to time you might find antennae or round fingertips. Small mouths and specific heights are also commonplace. You got the idea by now: the physiology of aliens is pretty narrowed. This means it is overly specific, as if one knew that most of them look like this. But no one does, and definitely not as precise.

What our little but valuable experience tells us is that species develop their traits in accordance with their natural environment and general lifestyle. Predators usually dispose of claws, sharp teeth and are very agile. Think of sharks for example. They are submarine animals and, like many others, don't have limbs. Hence, they don't have claws either. Instead, they have fins and powerful tails, which facilitate their swift movement underwater. And as the apex predators of their environment, they benefit from two or three rows of teeth, more than enough to tear their prey. Land predators like tigers, lions, cheetahs or other felines, are all designed so that their feet can quickly move about their joints, allowing for rapid movement in flat environments. Additionally, claws are present in these animals, so that they can grab hold of their prey in pursuit and also help themselves in other activities (teeth have the same role). Lastly, air predators – great birds of prey like eagles, falcons and hawks – don't have limbs or teeth. However, they have an exceedingly aerodynamic configuration and strong wings. Their beaks have solid jaws and are able to spot pray from miles away thanks to their developed vision. Lower class predators also have similar traits, though not as pronounced.

Conversely, herbivores (the prey of carnivore predators) have other traits. They, too, have developed a motion mechanism of their own, as they desperately need it to survive amongst predators. But when it comes to teeth or claws, they are not as imposing. Their attributes lean more towards the flight capabilities, rather than fight in the case of predators[1].

These evolutionary traits have change with respect to the current needs of the species. When the ice age came, mammoths, saber-toothed cats, giant ground sloths and mastodons were the predominant mammals on Earth. Carnivore and herbivore features were still present, but external protection increased as conditions worsened. This way, all mammals grew more hair, and predators especially had bigger teeth and claws to be able to rip through the extra protective layers of prey. Perhaps the best instance where evolution could be felt was when early amphibians migrated towards land, developing various new traits as to make way for new species.

So, with all those taken into account, what should we expect of aliens? They would be species just like the ones on Earth, only living in different conditions than what we are accustomed to. It would only be logical for them to evolve in such a way that they can easily adapt to environmental conditions, while conquering their planet at the same time.

Conquering a planet is no easy feat, but it's what a species is meant to do in order to become dominant, and later intelligent. That's what we did. And remember that a couple thousand years ago we were constrained to small parts of the world, mostly the middle east (parts of Africa and Asia). Little by little, we managed to take hold of almost everything there is. While we still have plenty of space and time to evolve, I would say that this result after a few thousand years of modern humans is quite promising.

As a parenthesis to this physiology analysis, there is a fine line between intelligent and non-intelligent species. Many scientists have tried to branch the given dispute, including those at the Search for Extra-Terrestrial Intelligence (SETI) Institute, but none has come up with a clear distinction. Perspectives include the way of communication, the efficiency of living, some even the life-span and evolution time. We don't excel at any of those on a closer look, but out of all the billions of species to have ever roamed the Earth, humans alone are regarded as "intelligent", so it's got to be something more.

"The ability to develop and improve technology while integrating it within your own society" – this is the point of view of some researchers. To me, it seems a bit too specific, but nonetheless appropriate for the only "intelligent" species we know of. A description which is simpler and more encompassing can be extracted from famous author Yuval Noah Harari's book "Homo Deus". In it, he elaborates on why the difference between humans (or intelligent life for that matter) and other species can be reduced to the former being able to "collaborate efficiently and in large groups".

Whatever the exact definition of intelligent life, one thing is clear – it needs time to evolve that way. And for that to even occur, it is mandatory that the species is not wiped out in the process, and that it is also able to surmount other predators. If we were looking for any form of life, it might surprise us that the best-case-scenario (as of today's lack of evidence) is primordial, bacterial life. But if we wish to describe intelligent, developed life forms, we should also consider that it's almost impossible for them to be the only species in their environment. That is, unless they are so advanced that there is no longer any need for other species to ensure their survival. Anyway, at some point in their evolution, other species must have existed. Concomitantly, these "other species" must have become less dominant than the civilization, otherwise it wouldn't have been capable of evolving this much. We are left with two cases to analyze for intelligent life alone (we'll get back to earlier life forms in a bit).

In the first case, just like humans, this alien civilization is somewhat advanced (to a planetary or multi-planetary level at max). Because of this restriction, other species – so-called animals – have to be present in their habitats. Their number and appearance are irrelevant to us, but as fierce as some might get, none should have the potential to dethrone the main species.

The worst alternative would be for more intelligent species to exist in that same environment, as in the infant modern evolutionary phases[2] one will unarguably try to conquer the other. Only one can come on top. It would be possible for two or more species to thrive forever in harmony, but highly unlikely that they would all have a similar level of intelligence and be all considered intelligent, at least in the development boundaries imposed by this first case. For example, lions and tigers, both apex predators, can live together in the savanna without trying to get rid of the other, but imagine if some other intelligent species like humans, determined to come up on top, were to exist on the same planet... it would be debacle.

So, in this first case, it is safe to assume that the aliens we are looking for are the one intelligent and dominant species. Because they are still around Kardashev level 1, it is also a great bet that they still possess most of the traits that put them in such a high position. Humans haven't changed that much in the last thousands of years; any difference is barely noticeable. Same would go for them. Considering that they underwent a similar evolutionary process (evolving from their planet's "primates"), they needed have certain features of their ancestors which came on top of the food chain initially. Here is where it becomes complicated.

We settled that they need to be apex predators and that their characteristics are mostly those of their ancestors. Depending on their habitat, we can draw some conclusions. The aliens must have some method of interacting with the surroundings – feel, move and operate with objects. Thus, some form of limbs is a must. Those could range from short (like the T-Rex) to long (like horses or moose). It's important that not all of them are disproportionate. For instance, kangaroos have small arms, but long and strong legs. Humans present one of the greatest proportions when it comes to interaction with the environment, so we could expect something similar. Per general, longer limbs could prove better, but not necessarily more efficient. A decent conclusion would be something in between. Also regarding their limbs, so form of grabbing mechanism should have developed. Again, this could be fingers, suckers, claws or something adhesive.

A digestive system and metabolism similar to humans able to provide energy for longer periods at once should be weighed as well. A transport system for energy distribution is also essential at this evolutionary point. Here on Earth, evolved animals make use of sanguine and lymphatic vessels. Another compulsory characteristic is some form of information distribution throughout their bodies – we have incorporated nervous systems. There is a subbranch of vertebrates, however, called cephalochordates, which doesn't present any form of proper brain and the circulatory apparatus doesn't include a heart. Yet these strange creatures wouldn't be able to achieve such great potential in their actual form, so we can rule out this possibility. Hence, aliens of this type would need some sort of digestive, circulatory and nervous systems.

Lastly, to hold their equilibrium, a skeleton structure is required. Something which encases the organs and supports the muscles is vital for durable protection in harsh environments. The backbone seems to be compulsory, too, as there are little to no invertebrate apex predators on our planet (most are insects, arachnids, mollusks or crustaceans – which are generally very small, not allowing for cerebral development either). In the past, some sort of developed crustacean used to be the apex predator of its habitat[3], but it is still unlikely that anything alike would be able to grant "intelligence".

A respiratory system isn't imperative though, as energy can be procured and further distributed to the body's muscles in a distinct manner. Neither are the mouth, nor the eyes or ears. Still, some kind of spatial orientation mechanism should be met in intelligent aliens, maybe along the lines of echolocation or magnetoreception. Maybe antennas aren't unrealistic after all. Even so, very few species on Earth are completely blind or don't have eyes at all, so we could anticipate a visual reception mechanism from a decently advanced alien civilization.

Notice how the last paragraphs seemed overly specific. They actually aren't. It is only common sense to expect alien traits similar to what we have on our planet, because that's the sole piece of information we can rely on. Other specific characteristics might emerge, but we would not know of them unless we knew the exact conditions in which aliens live. Like the stripes on tigers or the spots on leopards, it's for the environment in which they thrive that they developed these particularities. Same goes for height and movement, where many other physical factors play a role (gravity, nature of the habitat and thermal properties of it, just to name a few). We can only estimate that the size of them cannot exceed a certain threshold, above which the bones wouldn't sustain the body's weight and inevitably collapse.

This isn't that specific if you'd ask me. There are only some wide margins we could theoretically set to abate the possibilities. And even those are only prima facie. But if that is broad, wait until we analyze the other case.

A second case would focus on an intelligent species, more evolved than the first, which has already reached the rank of interstellar civilization and above (Kardashev level 2+). There are no upper boundaries to this scenario, as we don't actually know how advanced can a civilization be. It wouldn't influence our discussion anyway. By this point, it became obvious that the species in question must be the apex predator of its environment. However, other species become irrelevant as we move up the evolutionary scale.

Since we are talking about an interstellar civilization, its technology is sufficiently developed to provide efficient energy and nourishment sources, so that extraterrestrial animals are no longer a need. Except they are. While aliens of this level are probably more than capable of sustaining their environment on their own, biodiversity is mean to assure this preservation naturally. It wouldn't make sense to get rid of it just because. This would significantly impact them, adding an unnecessary struggle to their almost perfect society. Plus, it wouldn't be ethical, but we leave this for later.

Trait-wise, it is more than challenging to make any predictions. Becoming interstellar means that the civilization should've had some time to evolve in its own system and become powerful enough to harness energy at high efficiency. Because of this interval, which can by no means be though of as short, aliens of the same species but in different environments and even on different planets or systems would begin to show distinct traits from their ancestors. Maybe they change in height and configuration due to the new conditions, or maybe some limbs and organs need change in order for the organism to adapt. These are but a small part of the assumptions we can make. It isn't sure, but close to it, that multi-planetary and inter-stellar species will have branched out by the point they reach that status. This means that subspecies of aliens will have come about, to the extent of them being unrecognizable or baring no resemblance to forebears or other subspecies. Not even broad boundaries can be set whatsoever.

When a Kardashev 2+ civilization is in focus, it is intuitive to put accent on their cognitive capabilities above all other predator-like traits. The technology and innovative methods of survival they have come up with are no match for secondary species. It's their intellect we are interested in. And the first thing that comes to mind is their brains, their nervous systems as a whole. But we are aware that bigger brains don't directly lead to higher degrees of intelligence. We know this from the animal world. Look at whales or elephants for instance. They are far from what we call intelligent life. Yet they do manifest far more advanced behavior that other animals. Then look at dolphins. These have a similar-sized brain to ours, and they are widely considered to be the smartest animals out there, from an evolutionary perspective. So, biggest brains don't mean the most intelligent species, but they sure give individuals higher chances of becoming such. Whales, elephants and dolphins have the biggest brains, but immediately after come humans.

We are not that far behind the first spot. What they do imply is more processing capacity, just like for computers. More information can be retained and, subsequently, more advanced behavioral practices can be developed. Currently, the capacity of the human brain is estimated to be of the order of petabytes, thousands of times larger than your home desktop. But computer power will keep growing for the foreseeable future. The famous Moore's law states that computer power is expected to double every 18 months[4]. However true this will prove to be in what's to come, it is just a matter of time before computers will outperform humans (more on that in a later chapter).

For an alien civilization to be able to compete with their own technology – or maybe not compete, but facilitate their evolution as individuals – their brain capacity is demanded to increase. A larger central segment of the nervous system would be the evident solution. This is the main reason why most aliens are portrayed as big-headed.

But there's also one other aspect to take into consideration: organic augmentation. This is a broader subject we'll closely cover later on, but the idea isn't that complicated. It's based on amplifying the abilities of living organisms using complementary technology. We are beginning to get the grasp of this area by trying to implant micro-chips in an attempt to augment brain capacity or certain other skills. For an advanced civilization, we can only assume this would be commonplace. Accordingly, brain expansion wouldn't be indispensable – they could very well work with implanted memory devices.

Very little can be said about how inter-stellar alien species might look like, and not even the most intuitive assumptions are to be fully trusted.

Reality is that we are clueless in the face of extraterrestrial species. We are baffled purely by their existence. But in spite of our not knowing whether they exist, how they could look and how could they behave, we keep looking. We are relentless. We don't even know if we should be looking in the first place, or whether it's best that we don't signal our presence. Who can tell what the apex predator has in mind?

"I'm not afraid of the dark I know. It's the dark I don't

that terrifies me."

Alan Dean Foster

The needle in the haystack

We've been asking ourselves this previous question for a long time, but decided to make our presence felt. After decades of fruitless observations, we made up our minds and chose to be brave, hopeful. It was the year 1977 when these decisions materialized in the form of two space probes, meant to signal the existence of our civilization. The names given to these probes were appropriate with their mission – Voyager 1 and 2.

Back then, a committee, led by American astronomer Carl Sagan, was designated to put together most relevant and definitory information about the human race. Everything was then assembled in the form of copper, gold-plated disks, mounted on board of the probes, aiming to convey the story of humankind.

These time capsules contain various aspects of our natural world – ocean waves, birds chirping, humming whales, thunder striking. Different types of music were also included, along with human voices spelling out greetings in many languages, babies crying and discourses of state leaders. Lastly, maps were etched on both disks, displaying the precise location of our Solar System with respect to 14 nearby pulsars[5].

Be it that some advanced alien civilization identifies one of these messengers, they will be able to decipher our encryptions and trace our current positions. I'm saying advanced as this would require some knowledge, especially for identifying the lonely probes. But once they do, it would be the most concrete evidence of external life. Of course, they would have to reach out to us, so it can be a two-way awareness, but we've got at least some time left until then. For the time being, the space probes have long escaped the Solar System (in 1990 they both surpassed the obit of Pluto) and are reaching out interstellar space. Voyager 1 was set to reach the closest star system, Proxima Centauri, in 40.000 years-time. Other attempts happened since, but none as courageous as the Voyager missions.

Yet it might not be just us who tried stating our presence. One peculiar signal was received by radio astronomers at SETI. It was coming from the constellation of Sagittarius and it consisted of an abnormally high intensity peak, 30 times stronger than the ambient noise. And it wasn't a short pulse. It took 72 seconds for the window to close, as scientists noted. The signal was bewildering, which is why, a few days later when it was analyzed, it was dubbed the "Wow! signal". Pretty wow indeed, but there's more to it.

"Tantalizing" as physicist Brian Cox described it, the signal wasn't like any other. It was registered at the 21 cm wavelength, which is the kind of electromagnetic waves emitted by the most fundamental of elements, hydrogen. SETI astronomers believe that an alien civilization might choose that exact, natural wavelength if they wanted to communicate, because of its universal applicability. There doesn't seem to be any mundane explanation: no satellite, military or local signal could be it. And no one is supposed to transmit on that frequency as it is specially reserved for radio astronomy.

Coincidentally, the Wow! signal was detected the same year we launched our space probes. Just a few weeks earlier in fact. But more than 45 years have passed and no signal has matched this one. "Since the Wow! signal we've heard nothing. We've searched in more detail, with better telescopes, more systematically... By no means exhaustingly, by the way, I think we can do much more! But we've searched and we've heard nothing..." Cox put it in a BBC interview.

We are still ransacking for hints of alien presence, but it is a bit disappointing that no solid proof has come to light. However, in the eyes of many, this is no reason to give up. Others are on a whole different page. Regretted physicist and cosmologist Stephen Hawking held a much more paranoid view on extraterrestrial intelligence and repeatedly warned about the dangers of humankind meeting alien civilizations. He wasn't that big of a fan of SETI's aim, but acknowledged that it is most likely that we will find just bacterial life, microorganisms at most.

This last claim is generally agreed upon by the scientific community. Advanced civilizations are, too, a possibility, but from our experience, not that likely. So, for the most part, astrobiologists are looking for biosignatures as they are called. Anything from cells to fossils to mineral traces might be clues to the existence of life. Though it is way tougher than it sounds. There are many challenges to be overcome. One of them is the low quality of our instruments. They are quite modern one might argue, but not state-of-the-art. And in the context of microscopic organisms, state-of-the-art technology is required. No exceptions. And it might take a while before this is achieved.

A bit discouraging really, but there are a couple other methods of extracting promising information by investigating potential worlds that might be capable of sustaining life as we know it.

Life signs?

What do I mean by "life as we know it"? This question is the foreground of all other assumptions made by scientists in their search for alien life, so it is worth analyzing in detail.

Organic matter is behind all living organisms and the fundamental processes which make them possible have one common denominator: carbon. This element is the 15th most abundant element in the Earth's crust. It might not seem impressive at first glance – just another element in the periodic table, not even that common, how come it is so important? But carbon has something special to it, something no other element presents: it is capable of forming stable bonds with numerous other elements. This is due to the four valence bonds it allows simultaneously (a bit of chemistry here). Because of it, carbon can give rise to a vast number of compounds, tens of millions that is. Those consists of chains of different lengths with various properties, including proteins, nucleic acids, lipids and carbohydrates, the very building blocks of life. So, thanks to its chemical and physical properties, astrobiologists are almost convinced that if life were to exist elsewhere in the universe, it has to be carbon-based.

Some speculate that silicon would be an adequate substitute for carbon as the backbone of biological molecules. This is because of their bonding similarities. But silicon generally arranges in the form of crystal lattices rather than long chains, which don't recombine in a manner that would allegedly support lifelike processes, like carbon does.

An additional requirement arises: these molecules must go through a series of reactions that give way to living matter. For that to occur, a suitable environment is needed to support the chemical compounds. Water is one such medium. But it's the type of medium that matters and enables chains to interact with one another, not water itself. However, unlike other similar environments, water is relatively stable, giving it a thermal advantage that favorizes the formation of life.

If those basic conditions are met, then life has the opportunity to thrive (if it happens to emerge). Earth benefits from a large quantity of water and also a decent amount of carbon. Yet, in the vastness of the universe, it isn't surprising that many places have similar conditions, even if the rules are somewhat strict[6].

Evidently, we began looking for other planets which might sustain life. As our Solar System doesn't present too many options, our main focus were other star systems. Most of them are just stars, sometimes orbiting other stars... not much potential for life sustainability if you'd ask me. But then there are those lucky few which also have planets orbiting them (we refer to them as exoplanets, because they are outside our system).

To date, a little over 5000 exoplanets have been identified, but the total number exceeds that by a lot. Still, even among those ones we were able to highlight some quite auspicious. We have dedicated part of our most advanced apparatus to the discovery and thorough study of exoplanets. The brand-new James Webb Space Telescope is the most fit for this job. By examining atmospheric features alongside with other important physical characteristics, these instruments were able to find exoplanets with traits comparable to

Earth's. There is even a scale called Earth Similarity Index, ESI for short, which describes how akin to our planet are those candidates. Outstandingly, some of them have better life conditions than our own! Yet nothing... But even if those planets don't host extraterrestrial life, they might end up being novel habitats for future human generations.

Strictly for alien life though, those exoplanets might not be our quickest solution. So, where else to look? I said that the Solar System doesn't present too many options. None similar to Earth, that's correct. Notwithstanding the lack of such worlds, there are others that have fairly appropriate conditions. This means that even if there aren't any (apart from Earth) with the "right" morphology, alternate means of sustaining life are present in a few places. You know, before adventuring out in the uncharted cosmos, it might be worthwhile looking in our "backyard".

One moon of planet Jupiter, called Europa, is perhaps the most intriguing from this perspective. On the surface, there's nothing but ice. It's a frozen satellite of sorts. But beneath a 25-kilometer sheet, a liquid ocean rests undisturbed. And not any ocean. One that is almost 10 times deeper than Earth's, containing as much as 3 time the total volume of water. In liquid form! That's amazing. And there's more to it. You may say that water isn't enough – you would be spot on. But Jupiter's attraction works towards life's advantage. It provides a reliable source of mechanical energy, essential for any organic process. Besides, this might also allow water to interact with the soil below it, giving rise to thermal vents, like those on Earth. In turn, these would provide not just the building blocks for life (carbonic structures), but also the energy for life. Spacecrafts are continually monitoring the activity of this satellite, revealing game-changer evidence of its fine-tuning.

Saturn is surrounded by similar-fashioned worlds. There is this one, small, icy moon, named Enceladus. Unimpressive at sight, but there's more than meets the eye. Just like Europa, under the shell of ice, there seems to be a salty, global ocean, heated by the contact with the rock at the bottom. Unlike Jupiter's moon though, the Cassini spacecraft (that's been investigating Saturn and its moons for a while) noticed geysers coming out Enceladus. By analyzing their composition, molecular hydrogen was identified. And this element, as banal as it is, is also the basic food source for microbes here on Earth. Maybe other essential elements are waiting to be found...

The biggest satellite of Saturn is, too, a potentially life-sustaining environment. Also known as Titan, this body twice the size of the Moon has a thick and dense atmosphere, composed of nitrogen, methane and many other organic molecules. Because of it, lakes and seas of methane were able to take shape at the surface of the moon. Now, liquid methane isn't as supportive as water, but in the proper conditions, paired with the organic materials in the atmosphere, it might be enough to sustain some lifeform.

This seems too good to be true. But they're here, right next to us, not light-years away. Possible life-hosting worlds, full of mystery. We just can't get our hands on too much yet, not until further exploration. And I strongly agree with Brian Cox here – we can do much better than that!

The problem we keep bumping into, apart from the very existence, is that even if the answer to this first question is affirmative, we would know nothing about extraterrestrial life. We've barely got hindsight to our advantage (bacteria and primordial organisms that is) and can just guess how aliens out there might function. We couldn't tell until we saw them.

In my eyes, this is the best example of the black swan theory in action. This metaphor of sorts – black swan or black swan event – refers to an encounter so far out of experience that one can't even imagine it until it happens: if you were a European farmer and have only seen white swans your entire life (many of them really), you would be astonished when a black swan appears on board of a ship returning from Australia, as you couldn't have even imagined those exist before you saw one with your own eyes.

It is this consistency between aliens and black swans that amazes me. We are mere farmers in universe's plan... having no idea when a black, or pink, or whatever swan may appear from somewhere not nearly as close as Australia...

This led many to develop a sense of modesty and consolation. But then some are confident that black swans exist. And not only do they exist, but we have also come across them at many points throughout history!

Extraterrestrial encounters

The interesting aspect which differentiates alien conspiracies from all the others is its universal applicability. No matter the context, neither political, nor social, nor economical, the existence of extraterrestrial life is a never-ending subject. Not being restrained by any circumstances, it has given rise to tons of opinions over time. And because there is no clear evidence to prove it wrong and we are arguably not advanced enough to express ourselves with certitude, it would be best to view the issue from all perspectives.

The debate has been going on for centuries. Millenia even. And while there is no definite proof for the actual answer, we are overflown by the ongoing question. Long ago there were ancient drawings. Then texts people have been trying to decipher. Nowadays, with the latest developments in technology, it became facile to communicate globally and provide others with information at the touch of a button. One consequential and evident effect was the staggering increase in the engrossment on this topic. In turn, this only amplified the conflict between supporters of the two sides.

So then, the query remains – who is right? On a collective level, we couldn't pronounce ourselves. So far, there is insufficient evidence to tilt the balance on one side or the other. Individually, however, opinions are more extreme (though summing up to a relative equilibrium of the balance). And what better evidence could one bring to strengthen one's argument than straight up alien interactions.

The entire idea of aliens watching over us or surveilling and analyzing our every move is as fuzzy as a puff of smoke. But on a general note, everything comes down to close encounters, be they social or just visual. Flying saucers, sky signals or down-to-earth manifestations all fall in here. The press is overwhelmed with them, so are the people, to the point they are fed up. And among the thousands of outrageous news, it is unclear what one could or should believe.

It was first just ocular witnesses. Weird-looking lights in the sky people couldn't find an explanation for. Most of them ended up being just natural phenomena or optical illusions. Then, what people recalled to be unknown spaceships roaming the skies, later called Unidentified Flying Objects (UFOs), were proved to be manmade devices of military or astronomical origin. The wide majority were just video hoaxes lucky enough to make it to the eyes of the masses. Alien abductions used to be a thing as well, but they aren't receiving as much attention as they once did. Crop circles and other eerie shapes on the ground were also impressive at first, but they too have fallen out of favor with most.

People began feeling something sketchy about all those alien-related news, disregarding them almost completely and focusing on certainties. And they are entitled to do so. Fool me once, shame on you, fool me twice, shame on me, as the saying goes. But maybe, just maybe, there is some truth in all the mess. There are still isolated events for which no justification has been found whatsoever. Even though they are few, it's enough to raise question marks and startle conflicts.

Saving the best for last, there are a couple of individual, presumed encounters with extraterrestrial technology, which were granted way more attention than any others. Those are the blurriest and, at the same time, the most dispute inciting ones out there – the source of all evil. Probably the most famous alien encounter was the one at Roswell. It all started with the collection of some sort of debris by the US Army Air Forces officers. They later stated in the press that the recovered object was an uncanny flying disc. When they retracted their statement, claiming that the crashed object was identified to be a conventional weather balloon, it was already too late. People have read the astounding news, and with a preposition like this coming from the military, they were holding onto it. The subsequent retraction only made matters worse, giving birth to a whole new conspiracy theory about the Government hiding precious extraterrestrial technology from us. Speculations also included alien lifeforms and characteristic equipment. Why would they do such a thing?

Before deciding to go to any protest, keep in mind how it all started. I repeat myself, but there is no certitude in any of those cases. Moreover, 99% of them already proved to be misleading. Conspiracies are all over the place, some more persuasive than others, though more often than not they start either from a misconception, or from the wish to gain adepts, or both. But who am I to judge? Maybe something is being held secret and maybe they have their reasons... But as Stephen Hawking put it, covering up knowledge of aliens for all those years would be the best job ever done...

Like Hawking, many scientists are skeptical about these conspiracies and about the authenticity of evidence regarding alien presence per general. That doesn't mean that scientists don't believe in aliens. If anything, scientists are the most determined to figure out the truth behind it all. The distinction between scientists and conspiracy theorists is that scientists start from utmost precariousness and fill in the blanks, working their way up to certitude, while the latter begin with the conclusion and work downwards towards the foundation.

"The Science Delusion is the belief that science already

understands the nature of reality in principle

leaving only the details to be filled in"

Rupert Sheldrake

And while scientists are aware of the skepticism a topic like this one demands, there are a number (a small number, but still) of phenomena they can't clearly explain and for which they don't rule out the possibility of extraterrestrial influence.

The most widely disputed discovery was that of Oumuamua. You may look it up if you want, but the concept is simple. There was this mysterious object, very elongated in shape, which was observed in 2017, weeks after its closest approach to the Sun. It was one of the many identified at the Hawaii Observatory, but the way it was shaped and how it interacted with the surrounding medium made it stand out. The details didn't seem to add up. No matter how hard astronomers tried to scientifically describe it, they didn't succeed. It was as if it didn't belong to this world. Unless it really didn't...

First it was the shape of it – a cigar-like object (from our perspective), more elongated than any asteroid ever seen before. The flattest asteroids have an axis ratio of around 1 to 3, this one had more than 1 to 10[7]. Then the smooth movement, a bit odd for an outside body hovering near the Sun. Non-gravitational of nature, as studies indicated, suggesting a possible radiation influence. This reasoning led to a pancake-like model, only seen from a side[8]. Also, it didn't leave any trace like close-to-the-Sun trajectories usually involve. The logical explanation would be that Oumuamua was just a remnant of a disintegrated rogue comet, as some proposed, but that too wouldn't agree with the flattened shape. It was by all means peculiar, for which rationale scientists seriously considered the probability of it being an object of alien origins.

It presented similarities to some of our spacecrafts. Space sails that is, those light-weighed objects which rely on solar radiation to gain thrust. That's not to say that it isn't something more complex. Professor and author Avi Loeb examines the issue more meticulously in his book "Extraterrestrial". But there is no consensus in the scientific community. The only way to settle the argument is by studying it up close. And unfortunately,

Oumuamua is long gone, following its apparently hyperbolic trajectory, receding into the interstellar medium.

However, there is one initiative, project Lyra (named after the constellation it seemed to have come from), which aims to reach the object in a matter of decades and have a closer look. But it needs funding. As Loeb implied, gravitational waves required billions of dollars in order to be discovered; dark matter consumes hundreds of millions, but there the results are inconclusive still; every major discovery takes effort, but also money to be made. We have to invest in these studies, otherwise we shouldn't expect grand results. But for the time being, we resume ourselves to what we know. We are not seeking to exit our comfort zone, so to say, or we aren't willing to take the risks needed for revelatory discoveries of any kind – Loeb also said: "extraordinary conservatism leads to extraordinary ignorance".

An even with all the effort and the money in the world, it would be foolish to await immediate results. Time is perhaps the most sculpting factor of any study. And for this great of an issue, we have focused too little too late. Though we couldn't have done it any faster, so maybe the right way to put it would be too little too early. On a cosmic timeframe, our civilization is still in its infancy, let alone the last hundred years in which we began seriously looking for answers. "It's like stepping on the grass and claiming that grass doesn't grow", as Loeb put it. You must stay days in a row to notice the difference. And even then, nothing special might show up...

So, in all fairness, there are lots of ingredients needed to gain the smallest bit of knowledge. But that's the sole method we could single out that grain of potential truth among all the fake news. Yet besides knuckling down to researching what might come, many have tried to investigate the past – maybe it's more insightful than the present?

Ancient aliens

From what we've witnessed in latest years, almost all encounters described the visuals and the dynamics of the supposed alien objects, but not much more. In fact, it is very little information for us to deduce anything which might come in handy. In contrast, studying the past has revealed a great many possible descriptions, tracing back to the most primitive civilizations.

Archaeological discoveries, along with ancient manuscripts and drawings are perpetually analyzed and deciphered, even as you are reading. I'm not talking about how the pyramids were constructed by aliens, nor that they raised or destroyed entire cities and cultures. It's rather about how actual depictions and codices of early humans might reward us with some insight on both the aspect and behavior of supernatural beings (unlike what we have today – almost nothing).

The ideas being put forward in this subchapter are obviously very subjective. That's about as subjective as it gets, really. But allow me to reiterate that if it weren't for this freedom of interpretation, these historical findings wouldn't receive too much attention. Maybe even the Holy Books wouldn't be as prized as they are and religious division would reach unnerving levels. So, this is not to offend anyone. But frankly it is astonishingly difficult for modern society to reveal records from thousands of years ago (if you can call them that). It is only natural that various opinions emerged. And in the context of guessing alien behavior and practices –

because that's what we do, we guess – such opinions may prove relevant.

The first time in history when we can account for civilization is the Sumerians' age. Entire other cultures and empires succeeded, but Sumerians are the backbone of it all, the root of the civilization tree. And the reason why they matter in our search for extraterrestrial presence is that many of their records were found to relate to their gods. Okay but Egyptians had their gods too, so did the Greeks, the Romans, the Indus civilization; why are the Sumerians special?

Maybe they're not. It's just that their inscriptions are easier to decode that those of others[9], and their gods are somewhat atypical. Not that they wouldn't present similarities to other civilizations', but they are, from what historians inferred, closer to humans. They did present supernatural abilities, only not as unbelievable as others.

These deities that go by the name of Anunnaki were more connected to humans, being involved in many of their activities. They also bared resemblance: humanoid in shape, with similar proportions, but much taller (2-3 times a normal human). They were always portrayed with something on their heads. The Sumerians were also inspired by those gods in their quest for development, both social and technological. You might have heard that Romans made use of aqueducts, irrigation systems and built durable structures. Well, Sumerians were even more developed for their respective period. They were way ahead of their time. Apart from the complex writing system – cuneiform – and perspicacious methods of cooking, hunting and agriculture, they are also credited for pioneering mathematics, astronomy and astrology, inventing irrigation, starting the first schools, devising the first codes of law and also imparting time in days, hours, minutes and so on. That's quite something for 6000 BC.

The abundance of "inherited" knowledge and technology, and also their continuous apparitions in Sumerian texts and engravings pointed in one riveting direction: maybe those Anunnaki gods were more than just sources of divine inspiration early humans believed in. Maybe they truly interacted with them. Maybe they are the aliens we were looking for all along.

You notice how the whole discussion is filled to the brim with "maybe". And while nothing can change that as of now, it is intriguing to dig into those ancient records. The Anunnaki were more familiar to us, one could put it. They didn't have the head of an eagle or the body of a lion. According to the Sumerians, they were just humanoid beings with abilities way superior than theirs, that's why they were considered gods. But we may move back to the statement "any sufficiently advanced civilization is indistinguishable from God". See where this is going?

The Anunnaki appeared in other ancient civilizations as well. Akkadians, Babylonians and Assyrians also described strong connections with this sort of gods. Renowned author and journalist Jim Marrs further examines the linkage between these gods and aliens in some of his works. But as it goes, it isn't that much more than a speculation.

The disappearance of Anunnaki gods is a bit suspicious as well. We don't interact with them directly anymore. It was the Sumerians, the Akkadians, the Babylonians and the Assyrians which had interference. But then at some point they vanished. No more records of them. People have turned to other gods, leaving a blank space regarding the Anunnaki's withdrawal from humans' lives.

This gap leaves room for interpretation. If these deities were indeed extraterrestrial beings – not supreme entities as they are usually portrayed – their fading may be explicable. It could be that they lived here before us. Their knowledge was intricate and chose to share it with us. But maybe they were on the verge of extinction. After all, there are only a few mentioned in ancient texts and drawings. By giving out their knowledge to others, they would "assure" that it isn't lost. Another possibility is that they were sent here. A small number of Anunnaki, sufficient to handle us, and meant to help us develop and form the advanced society of today. Once their job was finished, they simply left, as benevolent as they had arrived. We wouldn't know how or where... Or maybe they were only curious. We could expect from a civilization like ours or even more advanced to be characterized by unquenchable curiosity. They might have just analyzed us and played with us until they got bored. Or maybe they had other interests...

Whatever it might be, their seemingly harmonic interaction with Sumerians and not only is, in our eyes, at least a bit weird. Even if we knew for sure, we couldn't justify their manners and sudden departure – we don't have enough records.

Another proposed idea which is continuously sparking controversies is one you come across in the Bible. In this case, the Christian divinity is interpreted by some as other-worldly beings interacting with us, humble humans. More specifically, it is this fragment in the book of Ezekiel:

> *"I looked, and I saw a windstorm coming out of the north—an immense cloud with flashing lightning and surrounded by brilliant light. The center of the fire looked like glowing metal, and in the fire was what looked like four living creatures. In appearance their form was human, but each of them had four faces and four wings. Their*

legs were straight; their feet were like those of a calf and gleamed like burnished bronze. Under their wings on their four sides they had human hands. All four of them had faces and wings, and the wings of one touched the wings of another. Each one went straight ahead; they did not turn as they moved. Their faces looked like this: Each of the four had the face of a human being, and on the right side each had the face of a lion, and on the left the face of an ox; each also had the face of an eagle. Such were their faces. They each had two wings spreading out upward, each wing touching that of the creature on either side; and each had two other wings covering its body. Each one went straight ahead. Wherever the spirit would go, they would go, without turning as they went. The appearance of the living creatures was like burning coals of fire or like torches. Fire moved back and forth among the creatures; it was bright, and lightning flashed out of it. The creatures sped back and forth like flashes of lightning. As I looked at the living creatures, I saw a wheel on the ground beside each creature with its four faces. This was the appearance and structure of the wheels: They sparkled like topaz, and all four looked alike. Each appeared to be made like a wheel intersecting a wheel. As they moved, they would go in any one of the four directions the creatures faced; the wheels did not change direction as the creatures went. Their rims were high and awesome, and all four rims were full of eyes all around. When the living creatures moved, the wheels beside them moved; and when the living creatures rose from the ground, the wheels also rose. Wherever the spirit would go, they would go, and the wheels would rise along with them, because the spirit of the living creatures was in the wheels. When the creatures moved, they also moved; when the creatures stood still, they also stood still; and when the creatures rose from the ground, the wheels rose along with them, because the spirit of the living creatures was in the wheels. Spread out above the heads of the living creatures was what looked something like a vault, sparkling like crystal, and awesome. Under the vault their wings were stretched out one toward the other, and each had two wings covering its body. When the creatures moved, I heard the sound of their wings, like the roar of rushing waters, like the voice of the Almighty, like the tumult of an army. When they stood still, they lowered their wings. Then there came a voice from above the vault over their heads as they stood with lowered wings. Above the vault over their heads was what looked like a throne of lapis lazuli, and high above on the throne was a figure like that of a man. I saw that from what appeared to be his waist up he looked like glowing metal, as if full of fire, and that from there down he looked like fire; and brilliant light surrounded him. Like the appearance of a rainbow in the clouds on a rainy day, so was the radiance around him. This was the appearance of the likeness of the glory of the LORD. When I saw it, I fell facedown, and I heard the voice of one speaking."

Ezekiel 1:4-27

It goes on, but the point is that these creatures of the Providence were like nothing ever encountered by humans. They did, however, present similar traits to us – "in appearance their form was human", "they had human hands", "Each of the four had the face of a human being". The more you look at it the more striking the similarities. This perspective is not to be considered a blasphemy, for it has no intention of dismissing God in any way. But there is this possibility that the angels have more to them than it is purely depicted in the Scripture.

Everyone is entitled to their belief. We cannot assert anything neither about Christian divinity, nor about Sumerian gods, nor any other religion there is. We may just look on the issue from different angles, such to form a more adequate view.

The alien mindset

It's curious how all of these historical descriptions of deities mostly paint them as friendly and loving. Beings that have enormous potential, that could terminate us at their will, yet they only mean good. Occasionally they are angered by us and reveal a fraction of their ultimate power, but that's an exception. If you think about it, there are close to no accurate descriptions of supernatural creatures wanting to harm us.

Still, the only think popular culture care about when it comes to aliens is their limitless abilities for destruction. That's odd. We have no clue what their intentions might be, we just assume they are bad from the very beginning. Let me state this again: we have no clue what they would want. We do not know how an advanced, intelligent species might think.

But we assume it's bad for us because we do know how one intelligent species thinks... that's us, humans. As friendly as one might regard us, we are the most brutal and anger-driven species on Earth. Whatever comes in our way we either destroy or use to our advantage. We kill animals for food, for materials, for clothes. We keep them in cages and make use of them as slaves for our well-being. We have, perhaps unknowingly as we like to lie to ourselves, established the most sordid dictatorship our planet has ever seen. And as our power increases, as the separation between species widens, we are more and more cruel. We increasingly believe that it is rightful of us to feast on their misery, while we are living more leisurely than ever before. Who put us in charge anyway? Was it the divinity? The aliens? The evolution? Most agree on at least one of those. But it is faulty to think of it like this. Both morally and developmentally. We don't see the harmful effects right away, but they may impede our mission in the long term.

As it went, if a lie is repeated enough times, you can't distinguish it from the truth. And the truth we are stuck with isn't the shiniest. We are hopeful that aliens would be more understanding than we are, but why would they be? If history has taught us anything is that whenever modern humans interfered with other species or human communities, it didn't end up pleasantly for the others. Stephen Hawking emphasized this, claiming that if, somehow, alien life exists in a conscious, intelligent form and not just microorganisms, it would be like Columbus and the Native Americans.

"If aliens visit us, the outcome would be much as when Columbus

landed in America, which didn't turn out well for the Native Americans.

We only have to look at ourselves to see how intelligent life

might develop into something we wouldn't want to meet."

In antithesis with the well-intentioned ancient aliens we considered previously, if we track our progress through history we are only left with the chaos. So, maybe it isn't indicated to let others know that we are here, waiting for the "inevitable fate". If they are anything like us, we should hide and expect the worst. That's why I often say that the most disturbing thing we could find on another planet wouldn't be aliens... it would be humans. The greatest discovery of mankind wouldn't necessarily be the most cheered, but by far the most shocking.

It isn't accurate to assume that aliens would think like us, though. On the contrary, we should expect an entirely different way of thinking. As discussed in the Great Filter segment, humans are in an intermediary phase, something of a technological adolescence. We are not infant, as we have progressed a fair bit from wooden tools and cave-living, but we are still far from achieving technological maturity. And as it happens with the human body, the effects of adolescence play a role in our overall journey. We are aspiring towards improvement, constantly and collectively putting loads of effort into conceiving new alternatives to ensure a better living. Breakthroughs in science, medicine, engineering, technology you name it, are at the order of the day. From this point of view, everything is awesome. Things are moving at their swiftest.

But with great power comes great responsibility. In adolescence, a teenager begins experiencing the unforgiving reality for the first time. It will seem that life got tougher overnight. It will surely pose an obstacle. But it's these first-hand experiences that help the teenager prepare for the adult life. If it weren't for them, he would be caught off-guard, completely unaware of what is going to hit him.

We are that teenager. That's not to say that we are totally unprepared. But we have got a few more obstacles to surmount before being ready. The problem is that these couple of obstacles are incredibly rough. We need some huge leaps to get over them. If we do manage to do it, however, our civilization will be unrecognizable. Many of these will be encountered throughout the rest of this book, but one that is truly paramount is cooperation. Letting our differences aside and learning to work with one another, aiming for the common good.

We know this is the way to go, but we aren't trying hard enough. Therefore, we don't succeed. And this messes up our Feng Shui. We get angry and rebel. And this rebelliousness turns against ourselves, leaving nothing good behind. We do not think beyond what's obvious and simply call the anger inside of us, anticipating that it will resolve the issue. Our actions are thus defined by this anger-driven behavior, by the rebelliousness of our still teenage character.

It is safe to say that any alien civilization that is capable of reaching us in the first place (and not just send random signals like we do) would be hundreds, if not thousands of technological-years more advanced[10]. For it to find us, then examine us and then also come to visit... we can only dream of such a thing. Their society would implicitly be more developed than ours. Collaboration is mandatory for a Kardashev level 2 or 3 civilization, so it might just be that they would've overcome those barriers, as we see them, by that point. And they would be, at least to some degree, aware of our status, so they will have done their research.

Physicist and futurist Michio Kaku made a nice analogy. Assume you are a squirrel in a forest, living peacefully, minding your own business. One day, you encounter two odd figures: a hunter and a developer. Both have their appropriate equipment: the hunter wears camouflage and has brought tools and guns, while the developer is dressed in a costume, wearing sunglasses and a suitcase. You don't know much about either, but the first scares you. You anticipate some crazy outcome in which the hunter is looking to shoot you and your kin. But let's be serious, who hunts for squirrels. He has no malicious intention, he may just want to give you some acorns, yet you think the worst of him. On the other hand, you don't pay too much attention to the developer. He means no harm, right? Otherwise why would he come unprepared? You couldn't be more wrong. He's got all the necessary arsenal to destroy you in that little suitcase of his. By just standing and "admiring" your habitat, he is actually planning to build an industrial complex in its place, wiping everything out in the process.

We are often tricked when in similar situations, and the deal with the aliens is a great fit. We see them like the hunter, evil as portrayed by nearly every source, when in fact we should be more afraid of the developer – our inner, dark self. While working to our sole advantage, we damage everything that surrounds us. Not that the hunter couldn't kill us, it's just not in his interest.

That raises another question: how can we know the interest of the hunter? We are not nearly as strong as him and have no idea what's going on in his mind. But even though the difference between a squirrel and a hunter is the same or lower than that between us and an advanced extraterrestrial civilization, we have the gift of intelligence. We have the ability to reckon the mindset of aliens, even if we don't resonate with them.

That's what everything in this chapter adds up to – predicting the aliens' way of thinking. There could be a number of reasons worth analyzing, but could, in my opinion, be ruled out from the start. By overcoming the so-called obstacles that we are currently facing on a daily basis, aliens will have reached a higher stage of consciousness. I am not to get involved in any kind of psychological debate, though I am pretty confident that they will much more focus on their common needs than on their pleasures and reputation. So, unless they are some sort of interstellar mercenaries, their motivation for reaching Earth shouldn't be anger-driven or status-driven. In other words, they wouldn't come here just to terminate us for the fun of it.

There must be some interest at the middle. In H.G. Wells' novel „War of the worlds", aliens weren't looking to kill us to entertain themselves, we were just in their way. They had some bigger quest and we were not irrelevant, but annoying, complicating things for them. That's why they began their killing spree. We should try to understand what kind of quest it might be.

For starters, they might want materials from us. They saw an abundance of certain minerals and useful resources on Earth and decided that they could make more use of them than we could. This may not justify killing us all, but it could be a reason for it. Though, if you think about it, it wouldn't make much sense. The universe is abundant with all elements, some rarer that others. At that level of development, practically any celestial body could be monopolized for its resources. Anything from asteroids, to planets could be mined for materials. Stars' energy could easily be harnessed, maybe even that of black holes. And what can't be found can be synthesized, however in smaller quantities. Plus, we would have nothing to offer them. Elements on Earth that we prize wouldn't be that useful for them. Maybe platinum-based elements or others with impressive physical properties would be a thing, but it's not worth coming all the way here to just get what little we have.

The only thing our planet has that's peculiar are the proper life-as-we-know-it sustaining conditions. Notwithstanding the fact that our planet isn't special in that regard either (there are billions of similar planets in the Milky Way alone), it would be illogical to travel all that long for colonizing an already exploited planet. Earth is, indeed, quite suitable for extraterrestrial life – we assume – but with the mayhem that's been going on lately, it slowly becomes impractical. Terraforming, as we call it here on Earth, would be a real possibility for them as well. We are aiming to terraform nearby planets before settling on others, so it might be an idea aliens would also share. But if every terraforming-allowing planet nearby would have already been fully exploited, then heading straight for habitable ones would be more favorable. So, if we suppose that all Goldilocks planets in their vicinity have been colonized and they are looking for further expansion, they might have a keen eye for Earth. This, too, is doubtful, as there are multiple other habitable planets in a decent range from Earth that to this day show no sign of alien presence. If that's the case, we would surely encounter it only some hundreds of years from now, at least. Artificial expansions could be a more viable solution too, but this depends on other aspects of their society.

If they've really run out of options, they might choose Earth. That would mean that they don't mind interfering with other species – intelligent species, that's important – and might take any necessary measure to assure their successful mission. In the extraordinary unlikelihood and unpredictability of this scenario, we would definitely not be ready in our current state.

The space farers of Star Trek had this general order called the Prime Directive. According to it, members of the Starfleet weren't allowed to interfere with the natural development of an alien civilization, such as to introduce technology, knowledge or other values before they were ready. This looks like an advanced civilization's principle, but whether one would obey something alike is debatable.

It is possible that aliens don't follow such strict rules, especially when talking about the survival of their species. That's one rule every species has in common: its individuals work in favor of its perpetuation. At the point when it's a life or death situation, we could be sure that desperate aliens would do everything in their power to survive. But so would we.

Their technology would be no match for ours (assuming they would invade right away), but as they are on the verge of extinction and, in the first case, are approaching the situation violently, it might not turn out as well for them. For an advanced civilization to get on the verge of extinction it means that either some other civilization did their best to obliterate them, or they made a significant blunder on their way, one that cost them almost everything. Anyway, they would be at a great loss for options. And in their desperation for survival, it would be more noxious to use violence. Humble humans might be a worthy adversary. Even if aliens come out victorious, it would be a pyrrhic victory[11].

Another reason for one not to invade our planet is that phase of technological adolescence we are in. It would be like parents would try to keep in check a stubborn teenager – no matter how hard they try, the result is based on the teenager's willingness to cooperate and adapt. And we are the most stubborn of teenagers. Our extraterrestrial superiors may try to "educate" us, maybe not wipe us out but use us to their advantage (much like we use cattle for example), but we will not allow it. At our stage, it would seem most logical to us to utilize the most powerful defense methods we have – nuclear bombs. They couldn't stop them and not harm the environment at the same time. And if they let them be, apart from causing notable damage to their forces, inevitable environmental defilement will follow. Yet again, a rock and a hard place.

But then there's this other possibility, the positive outcome in which aliens try to cooperate with us. If desperate, it would be a pleasant course of action for both sides, none would have to suffer and each will benefit – we from their knowledge and they from a suitable habitat. If not, curiosity and social development might be the cause for it. Either case works.

As the fictional Prime Directive implies, any act like these may alter the natural evolution of the lesser species. But it's not like we are ants or something. We do have some knowledge which helped us understand the way nature manifests. We have organized societies and devised moral principles. And we are trying, though not efficiently, to contribute to our civilization's development. We like to consider ourselves more apt than we are, but as arrogant as it may sound we are somewhat suited for this.

A harmonic coexistence of two intelligent species on one planet isn't impossible to maintain, especially at this point when both are evolved enough to understand certain aspects of reality. For one, it would be an enormous downgrade if it settled here definitely. If not, it would be just an interesting experience, like teaching quantum mechanics to kindergarten children (or worse). For the other, us, it may be the greatest gift of the universe – other than life, obviously. An inexhaustible spring of knowledge.

There would be no negative aspects in the whole picture. But some suggest that another factor may play a literally vital role. Even if aliens are well-intended and we find a way to communicate with them quite properly, they might unknowingly bring doom upon us by contaminating our planet with interplanetary material. This phenomenon known as interplanetary contamination is perhaps the most unexpected of them. An advanced civilization may have no issue fighting natural diseases, but anything out of the ordinary might overwhelm them. Same happens for us, and for the aliens. We could be infested by their plagues, but so can they from something as basic as common cold.

We turn back to our history, especially when the colonization of America occurred. Hawking said that it didn't turn out well for the natives, but it wasn't just the wars. It wasn't just the anger-driven motivation of European colonizers. Unknowingly, they brought infections endemic among them to the isolated people of America. Much of this was associated with respiratory tract infections, including smallpox, tuberculosis, measles, and influenza. That's how most of them perished, not wars and criminal intents on behalf of the colonizers.

One might say "ok, but colonizers weren't too friendly with the natives either, why would aliens be?". That's true, but apart from colonizers being humans going through the darkest ages (so even more uncooperative and irrational that we are today), they also didn't have many places to go. It was this wish to expand, imbued with curiosity which led to the discovery of the Americas. There was also Australia and so on, yes, but they had no idea about it either. By pure chance they found America; additionally, the outcome might have been just as bad. Aliens would have more knowledge about what they wish to explore: they would investigate in advance and be prepared for what is awaiting. That's why we can't precisely compare alien colonizers to European colonizers of the XV century.

When it comes to contamination, however, the truth might be just as ugly. It would be risky for both parts. But driving is risky too. It causes way more accidents than any other means of transport with one in about 80 people finding their end in a car crash. It is also estimated that on average, a person gets into a moderate to severe car crash once every 18-20 years. The odds are not in our favor, yet we still drive. And not just for the sake of it, but we need it. We are, more or less, relying on driving nowadays. This is also applicable to our human-alien encounter. Knowingly or not, they might put our existence in danger, and so could we put theirs. But history also thought us something else, more macabre I might say...

Colonization aside, wars have shaped our species' conquest for the planet. From the smallest tribes to the largest empires to ever thrive on Earth, some half of their known history summarized to just that: wars. Conflicts between the given form of organization and another were ubiquitous for all periods of time, since humans could be considered humans. You wouldn't see that anywhere else in the animal kingdom. But when it comes to us, the most evolved of animals as we like to consider ourselves, we tend to prove how unevolved we actually are. For the most part, if things didn't go according to plan, we would go straight for violence, we would declare war to our wrongdoers. There was nothing in between.

How absurd is that? People wouldn't try to figure out a peaceful way of solving things, war was already considered an adequate solution. Arguably, war was sometimes regarded as necessary in the respective context. But I highly doubt there was a shortage of ways for solving the issues. Then there were those wars that emerged simply because one or a group of maniac aristocrats wished to reaffirm their superiority. That's how the immeasurable vanity of the ones in charge caused extreme losses among the innocent people and terrible regress in society. Medieval times were filled with them, that's why they were later called the Dark Ages. In a time of plague and all sorts of societal problems, all there was left for were irrational wars which brought nothing but misery upon our civilization. Then the industrial revolution came and things seemed to be heading in the right direction, only for the most horrifying events to occur – yet again wars. But this time, things came with a twist and the wars spread across the entire globe. At such a scale, the two World Wars account for the death of more than 100 million people.

Wars are gladly not as frequent as they once were, but, unfortunately, they haven't made their leave from the stage of our society. We are left wondering when they will finally end. Or whether they will ever end. We can sure hope that they will, but nothing is certain at this point.

An advanced civilization that has overcome the trivial obstacles we deem paramount would, most certainly, be more rational than we could be in our wildest dreams. Building on that, it is more than probable for them to be calculate and rely on other, friendlier methods of fixing struggles. That's why we eliminated the possibility of an anger-driven invasion of our planet.

But what if they are in a dangerous position themselves? Not something to put them on the verge of extinction, but something that has the potential of doing so. I am not talking about a self-destruction event or a universal pandemic or other sort of scenario that could jeopardize their species. It is alien invasions we are talking after all. What if they happened to be not as lucky as we are, with the appropriate conditions and security from other invaders, such that another highly intelligent species decided to rival them?

Wars as we know them are induced by our planetary immaturity, something advanced civilizations would've long left behind as they achieved a state of "adulthood". But it is possible that even a Kardashev level 2 or 3 civilization hasn't yet reached galactic maturity. Their powers are unfathomable for us, but the immaturity might vibrate with our way of handling things. This raises a decent chance of them acting just like us when finding themselves in a troublesome position (on a much higher level).

So, the interstellar or even intergalactic war scenario that is broken from the realm of Science-Fiction could be something we will face sooner or later. In this case, an alien invasion wouldn't need to be anger-driven as we are puny compared to the other powers they face, we pose no threat and are practically useless. But it might be a strategical invasion: that, too, would be just enough to end us all.

If aliens decide that our planet is not only suitable for them, but also has strategic position and characteristics in the greater war, the logical course of action would be to occupy it before anyone else does. Except we are already here... that wouldn't matter. Does the developer care how many squirrels lose their homes or die if he's got the green light for a profitable factory[12]? Of course he doesn't! He is working for the "greater cause", even though this notion is a hollow one.

We would be at the aliens' disposal, whether we want it or not. If we comply, maybe we could be of some use to them and they would threat us like slaves. But maybe we are utterly useless and we would only represent a pain for them, consuming the resources and interfering with their work while doing nothing to their advantage. What excuse would we have not to be wiped out? If we don't comply and just act like a thorn in their flesh, things might unwind much faster...

On the brighter side of things, this is unlikely to take place in the foreseeable future. Considering that in the narrow but existing time when we looked out for life signs we haven't found anything yet, we could see ourselves as pretty safe. It is true that estimating whether life is out there in the corners of the cosmos is problematic – we don't have sophisticated detection methods, we misinterpret signals and we see most of the universe thousands, millions or even billions of years into the past. In that time, life might have evolved and become clearly superior, but if that's the case, they didn't fancy coming too close to us. They might be relaxing in the distance, keeping a low profile, not being as eager as we think when it comes to spreading throughout the universe. We couldn't tell if faster-than-light engines are something aliens have achieved, we don't even know if it's possible. Same applies for all of the futuristic technology they might benefit from. In theory, something like this could happen tomorrow. But for our silence, as far as we are concerned, the space is devoid of life and "Star Wars" isn't based on a real story.

Cooperation is key

As per general, everything can be performed better when collaboration has a say. Maybe not at first, but with time the positive effects show up. You might think that a project can be better done alone than with a group of people; that you can put more work into it and focus properly as opposed to being constantly distracted by the more ignorant members of your group. But if you pick your coworkers appropriately and give your group time to adjust to the working conditions, the group becomes a team, with everyone being aware of their role and the importance of the common aim.

From our understanding and experience, major objectives can only be achieved with great cooperation. People working together, placing their mission above the divergences that might arise. Though it needn't be people. It could be animals as well. An ant can't pick up a pebble on its own, but when other ants understand their friend's goal they help accomplish it. Wolves are a good example too. When hunting a larger prey, they go in packs. The pack spreads out and surrounds the prey, then they attack from all sides, yielding a much bigger success rate.

So, if both humans and most (if not all) animals achieve the highest goals only in the presence of fine cooperation, why wouldn't the aliens abide by the same principle? And since we are only interested in those aliens that might come and visit, no doubt that quality cooperation was at the center of their development. In the eventuality that their progress has also brought about ration, they shouldn't aim for invading us – this utmost goal that they surely wouldn't have any problem reaching is probably not our doom. Rather, their mission has a curiosity background. They probably want to find out how we behave in certain situations and how we adapt to the changing conditions of human nature. That would be pretty rational.

But those friendly visitors will have done the research. It would be the most logical thing to do: investigate first and just leave the field work for last. Some suggest that they are continuously doing that already. Either by powerful telescopes or undetectable satellites, they could be analyzing our every motion. Every action that we do could be monitored right now by extraterrestrial beings. Funny to think about it.

Some go as far as claiming that they also have individual delegates already doing part of the field work. Aliens disguised as humans, taking part in different events and interfering with our lives without us ever noticing. Shapeshifting or other trickeries could very well be common use, so we can't really tell. It sounds preposterous, but it might be the case. Anyway, no significant impact or direct consequences can be traced to aliens and blaming them for the misery in the world is foolish.

But maybe they want seek nothing but collaboration. They want to help and educate us for the bright future of space-faring civilizations. It could be that they prefer working remotely to openly stating their presence. Maybe they don't intend showing up one day in colossal machineries because they anticipate an irrational response from our not yet advanced species. It's not like in the movies, not for now at least.

For the sake of those scenarios, let's assume that remote interaction is not one alien species' sole attempt to communicate with us. Say that they do get here one day, after years or centuries of study, finally ready to get in touch. How would we respond? Again, cooperation sums it all up. If we opt for it, things might unfurl peacefully and to the benefit of both sides. If we choose violence instead, best case they leave as abruptly as they came, worst case... you know it already.

Were it only for the vicious case the chapter would be long over, but there's just a little left. We are hopefully smart enough to choose the peaceful way out. We wish to befriend them and learn from them. How would we go about it then? Unfortunately, no certain answer can be provided to this question (as nothing in this chapter really). Even those that work on the frontier have little idea in this particular area.

Field linguists may be designated for the job at first – see how the aliens think, how they perceive things and how they communicate. Easier said than done. Those professionals often face difficult situations when they have to interact with all sorts of languages never heard of and communicate with their speakers (tribes and other isolated, natural formations). And while sometimes they succeed and correctly interpret the means by which those people communicate, victory would be inherently difficult to attain when it comes to aliens.

We know people share similar interests wherever they are at and whatever the culture. But that's because they are all humans, all part of the same species. We have serious difficulty interpreting animal signals for example. Now just imagine a species from another planet, totally unbeknownst to humans. The advantage would be that they are more intelligent. They might have no trouble deciphering our languages and fully grasp our vision of the world. Child's play. To us, not very much so. We could just start with basic things like showing them simple objects. "Here is a red cube", then we show them a blue cube: "See, same shape different color", then different shame same color and so on. Basic things to see if they can make it out. But they most probably can. They might just be looking at us like parents do at babies showing them the toy car they just got.

Another method some thought of is by trying to live with them, not knowing anything about them but trying to figure it all out on the way. Babies have a much higher information retention capacity than grown-ups. That's how the nature goes. Their unadapted minds are struggling to perceive so many things at once, but they do a wonderful job at it. So, what if we grow alongside them. Or what if we allow babies to spend time in their company. Some even proposed cross-fostering[13], as crazy as it sounds. Not the most ethical of methods, but the others might work.

The problem is that even if one does work, its glory will probably be short-lived. We'd Neanderthals to them, completely parallel with their knowledge and practices. Anything they would like to teach us is perhaps a far cry for today's humans. Everything they do and conceive contorted. Thus, we might have luck for a while, beginner's luck that is. But there is a limit to our current understanding. A limit that may only move further with time and experience, something which we couldn't get right away from our guests. All in all, there will surely be a point where we wouldn't be able to comprehend, no matter what approach (peaceful one) we take.

The solution is inconclusive. We don't actually know how to interact with aliens, I guess we'll just live on and maybe sometime see. What's unquestionable is the need for cooperation. We'll have to work together to understand them, and even alongside them to acquire the full knack. Risky? Yes. Time-consuming? Yes. Effective? No idea. But we must try. That's our best option. And maybe someday, with their help and benevolence, we might understand the principles of alien life and technology.

Conclusion

In retrospective, there's not much we do know. We don't know if aliens exist first and foremost. For which reason we can't tell their intentions. And consequently, we are unable to predict any outcome or plan ahead. The entire chapter is filled with irreducible uncertainty.

"We tend to use knowledge as therapy."

Nassim Taleb

Unfortunately, there are no good therapists in this field. No piece of information that can assure us that there's nothing lurking in the dark preparing to strike. Nothing to signal any external presence and nothing to ascertain that we are right about any of our predictions regarding that. And we can't do much about that. If it's not meant to be then so be it. But if it is, we can put some effort into speeding up the process and eventually make the grand discovery everyone is looking for. Only then will we be able to make somewhat accurate guesses.

We might have to wait a while until then. After all, others may be enjoying the dark, not wanting to be discovered. Just closely surveying the others or not even that out of fear that an alien encounter isn't the most pleasant of experiences. A dark forest where everyone is afraid that others might strike on the first occasion[14].

We don't know if we should make the most of the dark too, but it seems like we are eager to be identified. Science is at a bit of a loss as well. Our best card in the game of reason is unavailing this round. On a spiritual plane, the Bible is abstract as always, with the best relations with extraterrestrials possibly presented as divine interventions throughout the entire Scripture and especially in the Book of Apocalypse.

With no other cards up our sleeve, we are forced to cheer up with the last resort: guessing. Whether aliens exist, whether they are hostile or whether they are hiding for our or their benefit... Our best guesses are quite positive. We don't seriously consider alien intervention as the source of our demise, as it's much more likely that something else is, more mundane perhaps. At the same time, we cannot provide evidence that something along the lines of an "alien invasion" will not occur in the future. There is still room for that possibility too. But adding it all up, this scenario is quite unlikely, even in its unprecedented unpredictability.

But maybe not knowing really isn't that bad. In the seemingly endless detective work towards identifying and predicting alien intelligence, we are determined to develop even further. Search even better and aspire towards greater goals. Learn about the universe and about ourselves. Maybe it's all for the better at last. We just have to keep looking...

"Searching for life we reveal answers about ourselves"

Asteroids, often coming in elongated shapes

Cuneiform writing (Sumer)

Artistic illustration of an alien invasion (the hostile case)

1. The fight or flight response is something that all animals have developed. In a given situation that could endanger the life of an individual, it is inclined towards one of those, which decides the outcome of the confrontation. Nature has it that trophic chains are established among the coexisting species, so that there are the primary producers (plants), the primary consumers (herbivores, which feed upon plants), followed by secondary, tertiary, quaternary consumers and so on. The latter are generally carnivores, culminating with the apex predators, at the top of the food chain. And depending on each animal's necessities, specific traits have emerged, which is why some are more inclined towards fighting and catching their prey, while others are suited for a more attentive lifestyle.
2. We could be deemed as intelligent, but at the same time we are not all that advanced (not even a level 1 on the evolutionary scale). We can see the heated relations between humans, members of the same species. Many times in history, discrimination played a fatal role, everything adding up to continuous heated debates between us, generating gaps and wounds that will not heal for a long time. Only imagine if we weren't the only intelligent species of this kind on Earth...
3. A long time ago, a giant crustacean by the name of Anomalocaris (literally "strange shrimp"), was considered the first form of a top predator in its habitat. One of the oldest known animals with compound eyes and pin-sharp vision, Anomalocaris was the hallmark of modern insects and crustaceans. It had formidable grasping claws, but no legs, being thus forced to swim in the open water of the

Cambrian seas.

4. In 1965, Gordon E. Moore, the co-founder of Intel, made this observation that became known as Moore's Law from which he deduced that the number of transistors in an integrated circuit doubles almost every 18-24 months. Though the law is somewhat still applicable to today's computers, we don't know how long it will keep up.
5. The values of measured frequencies from those bright pulsars were provided, so that an alien civilization could extrapolate the exact position of our Solar System from this set of values.
6. The first criterion for detecting potential candidates for harboring life are those exoplanets situated in the Goldilocks zone of their parent star. That is, where one could be positioned so to allow the presence of liquid water (between 0 and 100 degrees Celsius). Many factors need to be considered, including the star's properties, the planets reflectivity and atmospheric characteristics (pressure, absorption rates at different wavelengths and so on).
7. An ellipse (oval) is approximately the shape of many asteroids. Those are described by the major and minor axis of it, which give rise to these notions of elongation and axis ratio.
8. A space sail works by receiving and reflecting the light from the Sun, or any other light source for that matter, therefore gaining the momentum of the incoming radiation. It is useful for such a device to have as big of a sail (effective area) as possible. This led to the model of a somewhat flat object, like a pancake, with a large area but small, seemingly elongated lateral section.
9. Sumerians used a cuneiform writing system that has

already been decoded for the most part, as opposed to other civilizations' which we are still struggling to understand.

10. We are interested in the technological capabilities of such visitors, not the actual age of their species or that of life on their planet. Life on Earth is already billions of years old, but humans have appeared just some 200,000 years ago; technological significance had been gained only in the last 10,000 years. So none of that matters. We are talking about the technological development of aliens. It could be that they are very young but still more advanced than us, but they could also be older and suffered technological regression periods and not be that far ahead.
11. A Pyrrhic victory refers to the outcome of a fight where the victor has suffered so much damage that it can't be regarded as too much of a victory (for example if a 25,000 men army is only left with 100 people after defeating a 10,000 one).
12. The developer would wipe out the entire forest in the process, leaving not only the squirrel, but its entire species without a habitat. Still, profit is its priority and not the survival of some animals he needn't interfere with.
13. Cross-fostering refers to the biological, genetic process in which offspring are removed from their biological parents at birth and raised by surrogates, typically of a different species.
14. The "dark forest theory" suggests that there are many species out there in the depths of the cosmos, but none wishes to interact with any other. They are unaware of others and can only assume that they would do anything to ensure their survival, just like themselves, even if it implies killing. So, in an eventual encounter, it might just

be a matter of who shoots first.

II.3: Cyber attack

"We worried for decades about WMDs – Weapons of Mass Destruction. Now it is time to worry about a new kind of WMDs – Weapons of Mass Disruption."

John Mariotti

Humans are not the best they could be. "We all stumble in many ways" as one verse in the Bible went. Envy, pride, greed – everything that meets the eye is somewhat imbued with them. But it is for those sins that we just can't get rid of that we became good at something. Something we should not be proud of, but something we are really good at: destruction.

We didn't always excel at it, though our tendency to do so was always there. From the primordial ancestors that devised weapons in order to provide safety and ensure survival we've come a long way. Larger structures of people began forming and this led to a kind of instinctual rivalry between one another. Each strived to be better than the other. So much so that it didn't focus on their own development, but rather the gap between their development and their opposer's, viciously tricking them into wishing the worst for their fellow humans.

That's basically how conflicts arose, wars and other conflagrations startled by shifts in perspective or misbelief. And that's how we became motivated to come up with new, innovative and automatically more devastating weaponry. As years went by, we perfected ourselves in an unrecognizable way. And it just so happens that this book is packed with those methods.

And if it weren't for the last century, the book would be half as thick. Technology was the engine behind this destructive feat.

With it, not war, but peace became ominous. That's because whenever there was peace, people were in fact getting ready for a possible upcoming war. World War I was a war of trenches, of men holding rifles and shooting each other in open field. Then, there was a cooldown of 20 years, and by the time World War II broke out, there were ships, planes, tanks and other sophisticated machinery that didn't predominate in the previous war. The atomic bomb was the peak invention. Those only got better in the following decades, just in case that some major power attacked another. Nowadays, war comes in two variants: either full-out rage, or utter silence. We'll cover the second in this chapter. One particular branch of it – cyber war.

The silence in chaos

The world is a crazy place. No matter how hard one tries to keep calm, they cannot succeed indefinitely. You could lay in bed all day, pretending there's nothing you have to do, but you realize at some point that it can't go on any longer. It's one thing to break yourself apart from all the surrounding problems that are eagerly trying to catch you on the wrong foot, quite another to boldly face them. But grabbing the bull by the horns is what we are demanded to do.

At no point were humans fully protected from external issues (maybe babies are, but we won't count them in). We always had to get ourselves involved in all sorts of activities, some more annoying than others, to be able to proceed with our lives normally, as leisurely as possible. It's just that the issues change with the social context of the time.

On the one hand, you could argue that people have nowadays more responsibilities than ever. We have to go to school, work, pay taxes, commute between cities and workplaces, all while doing the other regular tasks that have always been present. Society isn't that helpful with that either. If anything, it may look like it's working against you most of the times. It's all a huge mess. A far from utopic environment in which people really do struggle to put bread on the table and not be exhausted at the end of each day. Simply put, chaos.

On the other hand, mathematics tells us something crucial. Just like in chaos theory, there's always a pattern. A societal self-organization that provides us with a solution to the exhaustion problem. In chaos, there's silence. And that silence is given by one modern term known as digitalization.

The word describes a more complex notion than it sounds like, but plainly put it refers to the conversion of information into a digital format. Almost everything is somewhat related and relying on digitalization, from radios to television, from toy cars to planes, from basic products to entire factories. Our lives are dependent of it. You could try to live one day without any of that, no big deal, but it would be almost impossible to adapt to the long term. Everything would suddenly become much harder. And we are often blinded by the stress and uncertainty in our lives that we fail to realize just how easy it is for us from the other points of view. We suffer from phobophobia, the fear of fears. We are afraid of being afraid. And we concentrate on what's causing the fear, the chaos, instead of seeing the glimmer of silence.

The chaos in silence

Now imagine you wake up one morning, you go wash yourself and head to the fridge to eat something before going to work. Only the food is no longer cold, some of it altered. You figure the fridge stopped working, so you eat what's left. Then you look at the watch and see how late it got. Weird thing your alarm didn't go off earlier. Whatever, you rush to your car hectically starting the engine only to see that the garage door doesn't open – maybe the batteries on your remote let you down once again. You rush to the metro station, but no timer is displayed. An abnormally high number of citizens just as bewildered as you are impatient on the trains that won't come. You end up running a couple of blocks to work and when you get there the keycard doesn't open the door. Having went through such trouble, you try calling your coworkers to open the door for you, but the phone isn't working either. You turn back and see a couple of them going through the same problems as you. At this point you start wondering what is actually going on.

Everything is normal otherwise, except for all technologies. You could barely say that the rest is normal because there is not much beside technologies, that's how dependent we are. You assume its not the Earth's magnetic field that went down, as it couldn't have happened overnight and with no prior warning. But what else could it be?

What is happening is surreal. But it could very well be the result of a cybernetic attack. The beginning of one to be more precise. It will only get more chaotic the longer the breach lasts. And it's all caused by a group of people sitting comfortably in their chairs, in front of a computer, disrupting entire systems and societies with one click. No one knows who controls it all, but people will immediately start pointing fingers. Even Governments will blame other countries if it gets out of hand, that's the easiest way to justify the inability to keep it under control.

When every single aspect of our society is interconnected so thoroughly, it is just like a Jenga tower – you can remove blocks from it without causing a collapse, but if you remove the table it all comes down instantly. That's how someone with enough manpower and expertise can crash the entire system at once.

With the whole range of technological benefits of today, we are stuck in this bubble of ours. Like a cage in which we are captive, but also secure. Isolated from the dangers, but unaware when a greater digital one may strike. And because it made it that high up the list, you can be sure that it would definitely not be serene. The chaos in silence. Sun Tzu put it best: "In the midst of chaos, there is also opportunity", yet it's not the pleasing kind.

The world behind the screen

Power grids, traffic systems, financial systems, they're all connected. Building blocks of our society. They make up the entirety of a nation's infrastructure. All electronic devices can be counted in as well; they are intertwined with digital data around the world, forming up a global network with limitless possibilities. That's why the chilling probability that anyone can disturb the network almost as easily as they access it is not negligible.

With enough preparation and a strong desire to provoke mayhem among people, one could break through the security precautions of any system and disarray it. That's how the notion of cyber terrorism appeared. It is for those tendentious individuals with hostile intents that aim at disrupting societies and manage to do so without lifting a finger from their computer. Cyber warfare could thus be regarded as one of the cheapest forms of destruction, but with formidable implications.

It wasn't long ago that this form of malicious interference came about: XXI century. And while it was expected, we only then realized how rogue things could get. It only takes a minor attack to hack into a system. With a bit of effort entire cities could be powered off. It can get to the point where critical infrastructure is targeted and shut down, with not only power stations, but also transport networks, hospitals, and even all personal data of any person "plugged into" the system (that's basically everyone).

Infrastructure could stop working entirely, but it's not infrastructure that's most exposed. It is those devices and appliances that provide us with the most convenience. Smartphones, speakers, microphones, thermostats you name it, those can be affected with little to no endeavor. Those that do not benefit from a whole lot of protection besides maybe a flimsy firewall or an antivirus. Now, don't imagine your toaster starting shooting bread at you or vacuum cleaners going mad and wanting to suck up everything in their path. But things could behave unusually. Clocks indicating weird times. TVs randomly turning on and off. Printers triggering without warning. Small things, but with destructive consequences. Think about your devices recording and filming everything you say and do in your house, or on the streets or anywhere else for that matter. Information that then reaches a bad intended cyber villain, sitting at the other end in his desk chair and drinking his packed juice.

Businesses are highly reliant on technology, too. Hence, they are often the target of digital wrongdoers. The issue is that even if the alterations are major, it may take months or years to discover the source of it. Even more so if the effects are minor, but slowly pile up to something disastrous. Because what is the point of screaming you're there if the point is to subtly exfiltrate something valuable. So, when confronting with smart, experienced adversaries, companies and organization are totally unaware that a serious breach occurred for extended periods of time.

"There are only two types of companies: those that have been hacked and those that will be."

Robert Mueller

More than 70% of hacks remain silent for a while. The only ones that can rapidly be detected are the ones that imply massive disruption at once, the ones that inflict maximum collateral damage. And there are no rules of conduct in cyber war, so it could very well be just that. Shutting down electrical apparatus or power grids would be noticed straight away, but that doesn't mean that it's easily fixable.

Apart from those, there are also the major ones that can hardly be spotted. Here comes in the important idea: the worst isn't achieved by cutting of the information stream, but by modifying it. And anything can be subject to change. Think of traffic lights for a second. If they are shut down in the entire city, people will figure it out, drive more cautious until the problem is solved. But by altering every traffic light, or even just a few, by a couple of seconds, people might not notice. Accidents will take place and only then someone may realize that the system was struck. Anything that is subject to change has the potential of causing disruption among the masses.

Financial systems can be shut down on a huge scale, spawning wrath as people can no longer make transactions or withdraw money. Panic will be all over the place. But at least it's the same for everyone in a way. It would be much worse for a criminal to alter account information, yielding millionaires at abject poverty and vice-versa. A more elusive alternative would be for one to "reward" oneself with an infinitesimal percentage of every transaction, say a thousand of a cent for each. Nothing distinguishable, but with the restless bank system, the one behind it all would make a fortune in a matter of days.

The worst-case scenario for a catastrophic cyber attack might not be parts of our infrastructure shutting down (like water filtration systems, power grids or financial systems), it might just be an advanced actor slowly guiding us in ways we cannot comprehend. And as cyber analysts emphasize, the worst part is that it doesn't take an entire organization or nation state to turn everything against you. Anyone with the necessary malware (which is available on a wide scale) has the potential of impeding or damaging information systems.

Salvation?

As attacks get progressively stronger and tougher to detect, the risks also increase. Groups with no borders started forming, learning and growing together either for the welfare of society, or for the grim. Official structures devised special departments for keeping the cyberspace secure, analysts and hackers to ensure protection of their private information. Other measures can be taken both by private companies or organizations and by individuals. Any basic security precaution taken is a step towards cybernetic safety (or should I say less of a risk of being breached). Accessing only trusted online sources, not putting too much personal data online, avoiding dodgy links and websites, keeping record of your preferably different passwords and periodically uploading your data onto memory sticks or some other external memory storage should be the rule of thumb when it comes to cyber protection. Quality antiviruses and trustworthy firewalls are also a means of increasing security.

Those are useful against small breaches and one should definitely consider them, but a to the more experienced mastermind it's a piece of cake. The thing is that experienced hackers aren't interested in an individual's private data only in exceptional cases. Rather they focus on greater structures which might provide them with viable results if successful. And as more and more of those cyber terrorists came to the surface these last few decades, the belief that they will turn our society upside down to their sole advantage kept drawing the eyes of the public.

"The beast I saw resembled a leopard, but had feet like those of a bear and a mouth like that of a lion. The dragon gave the beast his power and his throne and great authority. One of the heads of the beast seemed to have had a fatal wound, but the fatal wound had been healed. The whole world was filled with wonder and followed the beast. People worshiped the dragon because he had given authority to the beast, and they also worshiped the beast and asked, "Who is like the beast? Who can wage war against it?" The beast was given a mouth to utter proud words and blasphemies and to exercise its authority for forty-two months. It opened its mouth to blaspheme God, and to slander his name and his dwelling place and those who live in heaven. It was given power to wage war against God's holy people and to conquer them. And it was given authority over every tribe, people, language and nation."

Revelation 13:2-7

"Because of the signs it was given power to perform on behalf of the first beast, it deceived the inhabitants of the earth. It ordered them to

set up an image in honor of the beast who was wounded by the sword and yet lived. The second beast was given power to give breath to the image of the first beast, so that the image could speak and cause all who refused to worship the image to be killed. It also forced all people, great and small, rich and poor, free and slave, to receive a mark on their right hands or on their foreheads, so that they could not buy or sell unless they had the mark, which is the name of the beast or the number of its name. This calls for wisdom. Let the person who has insight calculate the number of the beast, for it is the number of a man. That number is 666."

Revelation 13:14-18

The Biblical Apocalypse may foreshadow a scenario just like this. Bountiful in metaphors and symbols, the verses of the Book of Revelation portray two beasts, both gaining authority over the people of Earth. They follow the beasts unmediated and fail to obey God's principles. They can't. They are absorbed by the beasts. They control them like puppets and force them to meet specific standards in order to be granted the right to live. There are no exceptions, every man and woman on the planet must do as the beasts say and bear their mark.

Similarities become less hidden. In case of a global cybernetic breach, the puppeteers would be the hackers, the ones behind computer screens and not the ones that climb on stages and hold speeches. An "underground" society with the ability of outweighing all of the opposition with just the push of a button (not really but close to it). They could very well impose strict measures that ensure their reigning over us and reward their adepts with more freedom than the others, freedom that would be much valuable in the given context. The numbers may only call for a symbolic meaning but if taken literally, 42 months is quite a long period for a worldwide cyber-attack. Though it's not ultimately outlandish. Breaches can last for years at a time. Bigger ones, those that critically impact the system, have a generally shorter lifespan, but if a great and experienced enough community sets out to attain world domination, three years and a half would be more than enough to cause the purest definition of chaos. Nations collapsing, people going mad and fighting one another, everyone struggling to survive as best as possible in those conditions instead of focusing on dethroning the cyber terrorist regime. Even if major breaches have a rough time from the beginning as everyone works towards finding a remedy, in the case of worldwide madness, the anger-driven humans will start blaming each other and act in the detriment of their supposed adversaries, when the real threat is far from their prejudices. This would make up for a decent head-start for the maleficent, an unfair advantage that might just be enough for global terror – a cyber apocalypse.

"And the smoke of their torment will rise for ever and ever. There will be no rest day or night for those who worship the beast and its image, or for anyone who receives the mark of its name."

Revelation 14:11

"And I saw what looked like a sea of glass glowing with fire and, standing beside the sea, those who had been victorious over the beast and its image and over the number of its name."

Revelation 15:2

But just like in the Bible, there is always hope. The Book of Revelation concludes the beasts' reign after angels straighten the minds of men, bringing them with their feet back on the ground. Those that joined were punished, condemned for working against their fellow humans. The others were finally released from what seemed to be an endless curse. But the beasts remained. They didn't fully vanish from the realm of people, sign of the potential cyber terrorists continuing their attempts of flipping the society on its head. And they will always be there. It is inevitable in a fully digitalized (or close to fully digitalized society). Many consider an attack of these proportion inevitable, too... So, who can save us from it? It must end at some point, question is who the rescuer angels are.

Throughout this chapter, you've stumbled upon "hack"-related terms every now and then. And maybe your thoughts immediately flew towards the dark side of things. But there is this common mistake people keep making, the misconception that hackers are always evil-intended. The term "hacking" does refer to gaining unauthorized access to a data system, but hackers themselves aren't always the threat. Often times, they are the salvation.

There are, indeed, a number of types of hostile hackers, those with criminal intents that want to either disrupt society, cause damage to certain organizations or act towards their own advantage and the detriment of others. Those could be deemed as black hat hackers. These are the cybercriminals, whistleblowers and the elite (the cream of the crop). They are generally experienced individuals activating in the cyberspace and hitting specific targets for their own benefit, the latter even devising new methods of cutting-edge cyber terror to overcome their opponents' defenses. Minor hackers are also a thing, annoying others for their amusement but getting little benefit from their actions. Their impact is also insignificant in the big picture.

Yet their counterparts are equally qualified. White hat, blue hat or red hat hackers are hired by companies and Governments to test and keep their systems secure, and sometimes also track down black hats. By preventing cybersecurity attacks, they ensure somewhat of a safety for their employers. They sometimes work on their own while keeping a low profile, not stating their presence openly but still helping with all sorts of breaches. The upper category of ethical hackers consists of nation sponsored individuals or communities that access other systems in order to monitor and prevent international threats. These are the top hunters that look for the juicy prey – international and peak-level cyber criminals, who should always be on the watch.

Ethical hackers are our only hope in case of a cyber apocalypse, in case something gets out of control. We've repeatedly met with alike situations in which a targeted hack got out of hand and caused massive breaches in the neighboring areas (cities or countries even). Just like a virus spreads out from a lab and starts a killing spree, even though it was merely created for research purposes. Those are a real danger, and only benevolent hackers can prevent a large scale "infection".

Deeds along the lines above can be seen in society from time to time. For instance, the undercover international community of hackers Anonymous, who people usually regard as having criminal intents, are actually working for the good. These Dow Jones coexist in an anarchic society (though pretty organized in the cyber domain), and form a sort of resistance, acting especially towards liberating the internet from dominating corporations and the violation of rights. So, perhaps there is some hope.

"The safer you are the more afraid you are..."

Conclusion

A terrifying doom in the area of cyber warfare is extremely unpredictable. Hackers are very volatile and new weaponry pops up into their arsenal out of nowhere. They always improvise and overcome even the most difficult obstacles our benefactors put in their way. On the other end, the protectors design new defense mechanisms at an ever-faster rate. For the past couple of years there has been a relentless fight between the good and the evil. And even though the good usually surmounted the evil, things might change very soon.

The alarming pace at which information technology progresses and at which digitalization takes over the entire world generates the perfect medium for hosting these kinds of issues. In the future, there will presumably be way more digital devices. Today, there are about 20 billion across the globe, but this number is expected to double in a decade or so. It only goes up from here. Electric and self-driving cars will gain ground in front of their fossil fuel-powered cousins while your home will be filled with all sorts of electronic devices, leaving only the bed and the chair aside (maybe even those will be automated for your comfort). We will be more tech-reliant than ever, but the Jenga tower principle still applies – once you remove the table, everything comes down crumbling.

Connecting the dots, we understand not only that some sort of massive cyber attack is imminent, but that it's closer than we initially expected. Many claim that it is nigh, and will happen this decade. That's wild! But as frightening as it sounds, the possibility that it will actually bring humanity to an end is minuscule. The anarchy will be there for sure, resulting in catastrophic losses both monetary, structurally and on a human life level. Our general progress as a civilization will plummet maybe centuries in the past. A few years would be enough to take the place of an additional Dark Age. But for us to go extinct... slim chance.

Then there's the fact that it's unprecedented – we don't know what to expect. It might not bring humanity to a close and we might be super prepared, but it's just never going to be enough. A wide scale cyber attack will still take us by surprise, as there is no way to tell when one will happen, given the unpredictability and experience of cyber criminals. But we might just be developed (and lucky) enough to bravely face one of the worst artificial disasters, and slowly but surely get over it. We can never be overly prepared, but if we wish to stand a chance, we have got to fight back and improve our methods just in case. And it will be the case...

II.4: Ecological disaster

"Beware of false knowledge; it is more dangerous than ignorance"

George Bernard Shaw

The illusion of knowledge is the most resilient trap in history. Thinking that you are right about something or that you are more entitled than others to decide upon something can prove terrible, even deadly at times. Going round in circles boasting that your perspective is righteous and no other rises to its level. Yet that's not the case. And who knows what's on the line. What you have enrolled in and what risk you are taking. Those aspects might as well be buried in the fine print and you'll have no way to tell what comes next. But you can stop right there.

You may want to analyze it from all angles first. See the other perspectives. Inquire whether yours is truly the most viable one. Plainly put, don't be delusional about anything. Because it's this illusion of knowledge, the false impression that you are correct regarding something even though there is close to no certitude in it – not being open-minded in the slightest and refusing to even consider the possibility that you are wrong.

And it's not only for intelligence that makes the script, it is also the social context, even surmounting intelligence in many cases. Some of the most pragmatic people have been fooled by the harsh society, luring them into this unescapable trap of false knowledge. The world's dictators weren't dumb people. They were just so manipulative and influential that at some point they began manipulating themselves, being tricked by their own deceit. That's how they spiraled down into this eternal vicious circle that only caused sorrow.

So, we shall not refer to stupidity as the sole lack of intellectual capabilities of an individual or a group of people (or the human race as a whole for that matter). Rather, all of those aspects should be considered, as they all play their part in our society. For better or for worse, they will always be there, shaping the way we think and act.

But there's little good the contributors of stupidity can bring. It is mostly bad omen. And as the greatest curse there is, one that we keep falling into, it can surely "reward" us with catastrophic results. For a civilization as evolved (or unevolved) like ours, this is the root of all evil. That's where the dreadful stuff is born. Many of the upcoming apocalypses are sealed in this Pandora's box of stupidity, which is unfortunately so easy to open. But for the time being, we'll take a closer look on one of the worst stupidity-based artificial apocalypses – the world of ecological disasters.

Craving for perfection

Evolution calls for development. There is no doubt that development and only that has brought us to the evolutionary stage where we're at. We didn't become more agile or rational and we definitely didn't learn too much about cooperation. But with the little that we did do throughout the years, we managed to improve our conditions to fit our lifestyle.

We got better at organizing our communities, at understanding the laws of nature and the principles that govern today's society are, for the most part, at their finest. We developed more suitable human environments and devised all different sorts of means to provide both safety and comfort. Workplaces are available for nearly everyone and people can strive together for the welfare of their families, while also developing their community. The way we handle things in general is much more effective than it was some hundred years ago.

It can't be said that we are living the dream, but looking back on the dark ages filled with misery that mankind had to go through, it isn't far from that. And the tree may branch out even further. It's hard to guess how fast it will grow, but it goes without saying that it will continue to do so for a while. That is, if we don't mess up and cut its branches in the meantime. But there's one branch that stands out, much bigger than other: the industrial sector.

Machinery, equipment, and supplies that are used to produce other goods make up the leaves of this branch. Anything from products to ores to energy fits in this great sector. And by no means is it bad that the industrial branch outpaces the twigs, it only supports our quest for progress. But whenever a branch gets significantly bigger than the others, it calls for more attention. Precautions must be taken so that its growth is controlled. Otherwise, is might break and fall down the tree, maybe snapping other twigs in its way. Even worse, if it gets too disproportional it may topple the entire tree. This would account for extraordinary losses; it would be a horrible regression from what we are familiar with.

We don't have to cut that branch from time to time for the others to catch up. Neither do we have to deprive it of leaves to remove some weight and balance out the tree. It requires only attentive care – not letting it get out of hand until the others get on the same footing. With that, things will calm down by themselves as development will open up new opportunities and we will learn how to take care of the branches.

The point is that we are at the steering wheel, and nothing bad should happen if we treat the situation accordingly. But we've had our faulty turns and sometimes we came anxiously close to the edge of the cliff. The tree nearly toppled because of our precariousness, our illusion of knowledge. And many times, it was because of this industrial branch that grew too little to soon and took us by surprise in our already bumpy pursuit of progress.

Haste is waste

As with all the areas humans put their fingerprint on, doing something hastily rarely leads to pleasant outcomes. Going over anything at a faster pace doesn't imply faster progress. Often times, the aspects that matter most escape you, leaving no room for reconsideration, thus missing the essence entirely. That's how the expression came to be. And the past provided us with plenty of instances where it testified. Ecological disasters are no exception to this empirical rule.

Energy was always a big deal. People always needed it, be it for transport, daily activities and chores or, lately, electronic devices and appliances. It became a must of the contemporary society. And as illustrated in the previous chapters, hell would break loose if it was taken from us. That's more than justified in the given context. We are used to taking it for granted, a need so basic that it is found in every corner of life. Yet it isn't as easy to procure as it seems.

Most of the world's population (though not all of it) has access to energy with relative ease, but we fail to realize what a daunting task supplying billions with on-demand energy is. Humanity was more or less forced to find better ways of generating energy. Fossil fuels, which were a loyal companion of ours for millennia, seem to be running low. The industrial revolution came with a spontaneous burst in the usage of fossil fuels, rendering them exiguous for people's needs. We turned to more feasible energy sources, renewables. The elements of nature became our new friends. Wind, water and earth[1] proved to be exceptional energy sources, but a few other, more obscure companions joined the race too. Among them there is this mogul: nuclear energy.

We didn't quite get it until the last century, but then amazing breakthroughs reshaped the world of physics, and subsequently that of energetics. With the invention of the atomic bomb, scientists revealed ways to generate energy via reactions of heavy, radioactive particles. Not 10 years had passed and the first nuclear power plant was opened in the former Soviet Union. The Soviets have been dominating the nuclear power stage for a while, but a couple of years before the fall of the Communist regime the most horrendous of ecological disasters unfolded.

In the night of 26th of April 1986, one of the greatest sources of energy in the Eastern Bloc suffered an error so massive that it could be felt across the continent. The Chernobyl power plant (present day Ukraine, close to the border with Belarus), workplace for thousands operated by four RBMK reactors[2] at its best. But that night, it so happened that the newly built reactor 4 of the power plant underwent an irreparable malfunction.

Many factors contributed to that, but we will not get into the details. It was a routine test that caused the breakdown. Slight defects in the apparatus led to an abnormal decrease in the reactor's power, but it was imposed that the test must go on, in spite of the operator's insistence on shutting it down until proper recovery. Computers advised the same, but in the end, it wasn't stalled. Every safety measure was disabled in an attempt to reach the designated test running power.

Bringing human error in the equation, the result is meant to be bad. That's what happened at Chernobyl. In the absence of the control factors, the core simply obeyed physics' laws and increased its reactivity exponentially. The power increased above the testing threshold, even above the operating norm, resulting in a general panic of the operators. The "panic button" was pressed and for a few seconds things seemed to calm down. If only they knew that this last resort had an enormous destructive potential.

The workers never encountered anything similar, if something went wrong they would just wait and proceed when safe. But this inspection was different. Way different. By calling this ultimate safety measure, all the controls should've dropped back down into the reactor and stop the reaction. And they did. Well, the first part. They all dropped down at once. But the cases in which the controls were contained acted in the detriment of all safety measures. The poor graphite further accelerated the reaction, the power skyrocketed and the cycle kept going for another couple of seconds until it was too much for the reactor.

The others were carrying out their business as usual, unaware that a routine test was going on. They just heard weird sounds of steam coming out and trepidations in the vicinity of the reactor. Next thing they knew, the hundred-ton lid shot out like a bullet. Oxygen was now free to enter the reactive core, and along with hydrogen and other reactants, half of the building blew up. Everything shredded to pieces, radioactive ones.

Intuitively, every soul rushed to stop the expansion of the already outstanding danger. Operators, technicians and firefighters alike worked together, cooperatively, to reduce the danger posed by the lack of cooperation. Many knew what they were getting themselves into. But the civilians didn't. Thousands died in a matter of weeks following the explosion due to radiation poisoning. Highly increased cancer rates were reported in the neighboring countries (Belarus, Poland, Romania and, obviously, Ukraine) in the years to come, especially at children. All the precautions weren't enough to prevent the death of hundreds of thousands.

An exclusion zone was established around the power plant, but the danger slowly faded away. This was mostly due to the so-called Sarcophagus sealing the remains of the former plant. Living conditions aren't much worse than in other cities, maybe even better than in highly polluted ones, but the city of Pripyat and many others nearby are just shadows of their glorious reflections. Since then, nuclear reactors were improved world-wide, hoping that a similar scenario will never happen again. By the looks of things, people have recovered since the Chernobyl disaster. Still, what is classified as the most devastating nuclear disaster in history could've had way more terrible consequences, thrusting the death toll like the reactor lid.

10 days following the initial explosion, scientists and engineers that were on the sight learned about a far greater threat. As the first explosion severely damaged the reactor's water-cooling system, it caused a large pool to form just under the molten core, which had gotten out of hand. It was slowly making its way down into the ground, soon to reach the water pool. If the nuclear core were to reach this water it would have caused what experts call a steam explosion.

The explosion was estimated to be roughly 5 Megatons of TNT in energy. Minute compared to today's nuclear devices, but it wasn't the effective explosion that threatened our welfare. After all, the area around it had pretty much been isolated. However, the steam released would've infected the sky with immense quantities of radiation, which would have then spread across Russia and the rest of Europe. While the equivalent in lives in unknown (thankfully), physicist Vassili Nesterenko and many others claimed that this explosion would have made Europe uninhabitable.

Think about it: an entire developed continent of 800 million people engulfed in a long-lasting radioactive cloud that would sooner rather than later suffer catastrophic losses because of radiation poisoning and increased cancer rates. Luckily, we didn't quite get there. A brave team of 3 workers at the plant signed up for the suicidal mission of closing the valves and stopping the water from spreading further, thereby averting the utmost massacre of humankind. They were told that the contaminated water would bring them to an end in a matter of days to weeks, but they've lived many years after, 2 of them being in well health even nowadays.

Behind the scenes of the Chernobyl disaster lie many other factors, better explained in Adam Higginbotham's "Midnight in Chernobyl". But for the purpose of this book, it is enough to just get the grasp of the full-scale danger that might've unwound before our eyes, were it not for the boldness of not 3, but hundreds of people who paid with their lives to save Europe from being butchered.

14 years after the horror in Ukraine took place, another kind of ecological disaster came about, just some hundreds of kilometers South. Actually, this one happened right in my home town, Baia Mare, Romania. But this time around, it wasn't related to the energy industry, but the mining industry.

Baia Mare had been a mining city for decades in a row before the incident so, unlike with nuclear power plants, the mining industry was decently developed and people knew what they were doing. For the most part. The pollution was intolerable in the late XX century – the air contaminated with lead, yellowish in color, and the life expectancy going 6 to 12 years below the national average. All that could be traced to the mining industry. In short, the conditions were unacceptable by the World Health Organization's (WHO) standards, and it could be seen and felt. But that was just the calm before the true storm came rushing in.

Then, one month into the new millennium, a particular mining company raised the citizens' spirits when claiming it had the ability to clean up the toxic tailings at the Baia Mare mine, which had begun to spread as toxic dust by the wind. They promised to deal with them and extract the remaining gold from them in addition. The process used was gold cyanidation.

The technique was successful at first, with all the necessary metals being extracted from the tailings. But then, the company decided to deposit the waste product in a dam near the city. The dam then contained 100,000 cubic meters of cyanide-contaminated water – 100 tons of cyanide in total. But as the dam was overloaded and the heavy snow that winter didn't help either, it gave in. All the toxic material was spilled over areas of farmland and then into the river.

It's time for another mind exercise. The lethal cyanide does for a human is about 1.5 mg/kg of body. Taking an average of 70 kg per adult and we extract a number of a billion people. That's how many people could be killed by the 100 tons of cyanide spilled into the rivers. Other estimates go up to a few billions, but our result is more than needed to understand the scale of this spill.

Fortunately, it didn't really interfere with humans on its way. It did however, interfere with marine life over the thousands of kilometers on its way to the sea. As a consequence, about 90% of the aquatic fauna perished, including some 20 endangered species. Millions of Romanians and Hungarians had their drinking supplies contaminated and the toxicity of the surrounding rivers rose to 700 to 1000 times the admitted level.

Many volunteers helped clean the waters after the spill, but it was too late for the fish. This event was considered the second worst ecological disaster after Chernobyl (in Europe and maybe even the entire world). You would think that people have learnt their lesson, but no: just five weeks later, 20,000 cubic meters of heavy metals contaminated the waters and, one year after that, another cyanide spill was deliberately produced a bit to the East.

This incident wasn't nearly as dangerous as what happened and could've happened at Chernobyl, but if something on a larger scale were to happen, intentionally or not, many millions could have real issues finding clean water supplies. Imagine if a cyanide spill occurred in the deepest lake in the world, lake Baikal, which makes up a fourth of the world's potable water supplies[3]. That would be calamitous.

Those mistakes are reparable, just like the Baia Mare cyanide spill was in the end, and everything was brought back to normal (if you could call it normal). They aren't very likely either and the outcomes wouldn't bring an absolute apocalypse upon us. Still, they happen from time to time. Human error is one of the most unpredictable aspects there are, and so is its destructive potential.

You might have gotten a poor impression about Eastern Europeans' working quality. I can't really judge you. But similar incidents happen all around the world. The nuclear accident at the Fukushima power plant, Japan, had exceedingly high safety measures, but an earthquake-tsunami duo proved devastating for the plant. The lives of many could've been in great danger if the threat hadn't been concealed as quickly. India has witnessed a terrible accident, too: a toxic cloud of methyl isocyanate gas escaped from a pesticide plant in the city of Bhopal, taking the lives of 15,000 residents in just a few hours.

Anything along those lines could theoretically be take to another level, such as to yield much heavier life losses and impact the surrounding environment for years. But even through mistakes like these, it's hard to achieve apocalyptic scales to be frank. Except it's not... There is one slight mistake, perhaps smaller than all the others so far, that could theoretically bring doom across the planet. That's the world of bioweapons for you.

Hidden in plain sight

You see them everywhere. They are thriving around you at all times. They lurk in the dark, but also in bright daylight. They are present on any surface and in any medium. The air you breathe, the water you drink, the phone you use and the pages of this book you're reading, they're full of them. Silent assassins hiding in the microscopic realm, too small for our eyes to perceive. And they are resistant, persistent and spread faster than bread crumbs, with the apparent sole purpose of causing suffering among living organisms.

Viruses, bacteria and fungi all fit the above description. Some of them are vital to the well-being of ecosystems, supporting much of life on Earth, whereas others really are here just to harm us. If you were to choose between living with them forever and them going all extinct, it would only seem logical to wish for the latter. But that would be a terrible miscalculation, resulting in losses greater than all viruses combined. Epidemiologist Tony Goldberg put it simply: "If all viruses suddenly disappeared, the world would be a wonderful place for about a day and a half, and then we'd all die – that's the bottom line". That's how important they are. But as with all things, there are a few exceptions, those that we are accustomed to. They are the deadly ones, the unseen arsenal of nature.

When we talk "bioweapons", we generally imply some sort of direct damaging method caused by biological agents and inflicted via a transmission mechanism. It can therefore be regarded as a two-piece projectile, like the cartridge with gunpowder and the actual bullet that is being shot out of the gun. The gunpowder would be considered the infection mechanism – i.e. what makes the bullet fly as the gun is fired, and the bullet would be the damage inflicting agent.

These agents can be anything from bacteria, viruses and fungi to different sorts of toxins and poisons. Anything that can be regarded as biological and can be utilized to kill other living organisms. Due to their nature, they are more often than not uncontrollable, unstoppable even. And because of this, they pretty much top the lists of weapons of mass destruction, right there next to nuclear weapons. However, the term isn't the most appropriate for this type of armament, as unlike the others, biological weapons are incapable of destruction of infrastructure, buildings or equipment, only the natural environment, including humans, animals and plants.

Similar to nuclear weaponry, bioweapons were banned worldwide (180 countries approximately as of latest years) by multiple treaties, the most notable of which being the Biological Weapons Convention. The member states were prohibited from developing, testing, producing, stockpiling, or deploying them. Bio warfare was practically forbidden. In spite of that, some countries (and possibly groups of "dedicated" individuals) continued to pursue their biological weaponry capabilities, thanks to the low expenses and ease of developing such armament.

Another aspect they share with nuclear weapons is that their destructive potential couldn't be accurately estimated if it weren't for the bruising experience we went through. Fortunately, not us, but the past was merciless. Nuclear weapons were only manufactured after World War II, less than a century ago, and their utter power took the lives of some hundreds of thousands. Bioweapons were here since the commence of warfare. Infecting and causing serious casualties among enemy personnel without clinching was quite a common practice even before the Common Era. But the true severity of bioweapons could only be seen in modern times. On one instance, the British troops intentionally passed blankets infected with smallpox virus to the Indians, causing a devastating epidemic. It wasn't that bad at first, but please consider the number of smallpox fatalities in its last 100 years of activity: 500 to 600 million! It was the most disastrous infection of them all (until 1980 when it was fully eradicated).

World War I met biological warfare on many occasions. The Germans were notorious for their countless attempts and successes of damaging Allied forces, directly or not, with the help of bioweapons. A famous attack took place in 1915 at the Osowiec fortress (now Poland) against the Russian Empire. 14 German battalions were sent to take over the outpost – 8-9 times the number of Russian defenders. Still, the Russian defenses were strategic, and the Germans opted for the usage of chlorine and bromine poisoned gas. They thought everyone was dead and that it was safe for them to take over, but their opponents didn't give up and fought to the last man, chasing away the German soldiers (that's where a bunch of machine guns came in handy). The losses were devastating on both sides, but Russians came out victorious.

The victory wasn't long-lived however, as almost all of them died in the following hours because of the poisoning. The event is known as the "Attack of the dead men", inspired by the decomposing, zombie-like bodies of the defenders upon being poisoned.

Japan wasn't shy of using bioweapons just a decade after signing the Geneva Protocol against them. It violated the treaty's ban when repeatedly making use of them on Chinese in WWII. They further continued their development, reportedly killing thousands of subjects in the testing process alone. No other uses have been reported since then. At least not in conventional warfare.

But it wasn't meant to be where the merits of bioweapons end. Besides other isolated attacks, a new form of war emerged around the year 2000. That's when the war of terror became the central focus of world's powers. An unknown enemy, just as invisible as the bioweapons they put to use and almost equally "infectious". This mold-like adversary profited from the cheapness and effectiveness of biological warfare and has made its presence felt.

The nations of the world have thus invested not in the development of their own biological warfare program, but in their defenses against it. Scientists all over the world work shoulder to shoulder trying to prevent any biological massacre from ever happening again. But the problem with this type of warfare, just like with cyber war, is that it's extraordinarily unpredictable and accessible at the same time.

As far as we know, a nation could be secretly developing cutting-edge bioweapons right now. It could very well be that not a nation, but a group of people or one mad scientist that's been working his whole life to conceive the most powerful bioweapon in existence have finally triumphed. At the same time, some random researchers whose aim is to merely study biological agents might be exhausted from all the work and, as a consequence of not drinking their coffee one morning, accidentally release doom upon mankind. There is no certainty in this field. We are left to take all the precautions we could think of and prepare for the worst.

Conclusion

"He called out in a loud voice to the four angels who had been given power to harm the land and the sea"

Revelation 7:2

"From the sky huge hailstones, each weighing about a hundred pounds, fell on people. And they cursed God on account of the plague of hail, because the plague was so terrible."

Revelation 16:21

"A third of the waters turned bitter, and many people died from the waters that had become bitter."

Revelation 8:11

Does this sound familiar? Of course it does. The entire chapter could be summarized to these three verses in the Book of Apocalypse.

But the Bible isn't very indicative of the outcomes of the wind-carried plagues and diseases. The whole point is that the people had to deal with them until eventually dying out. The hailstones are about as blurry as it gets. They couldn't originate from something else as there would be no connection to the raining plagues. Yet it seems to be totally independent of the presented hail whichever way you take it. Reasonable explanations would somehow relate it to severe structural damage, like what happened at Chernobyl when half the building was blown to radioactive fragments which workers then had to get rid of risking their lives. But it couldn't be said that the fragments really fell on people killing them, so the metaphoric meaning may be even deeper. The harmed "land and the sea" are quite descriptive of the outcomes of the two presented ecological disasters. Toxic fallout could significantly impact the environment, leading to the terrible plagues described afterwards.

Whatever the case, it's tough to anticipate any of the events covered in this chapter, but we could at least outline the possible casualties following such a scenario. Though even those would make up for a very broad sketch. Ecological disasters could happen spontaneously, at any time. Note that we didn't touch the other areas of environmental issues since they deserve a separate talk and are distinct from these by one essential aspect: these ones can't be predicted. We must take all measures, but it could still not be enough. Some rogue factor could randomly alter our equation and yield totally different results. Catastrophic results.

As for bioweapons, they are even less predictable than environmental disasters. Also, they could account for much greater death tolls and over shorter periods of time. They're not to be played with. Just like you don't give matches to little kids, bioweapons shouldn't have been given to us. Yet they were. In fact, we found them ourselves. And for a child it's just as fascinating how a match burns as it is to us how these pathogens and poisons can evolve. That's until you get burned – not so fun anymore.

Most of the times, the match burns out, maybe consuming the stick in the process. But on some occasions, it might come in contact with something flammable. By accident or intent, the object bursts into flames, lighting on fire nearby materials as well. It's pandemonium until the firefighters arrive. Chance has it that bio-firefighters aren't as experimented as regular ones. It will take a while until the weapon is deactivated or rendered futile. However, enabling the evasion of detection tools, enhancing the virulence and mutations and conferring resistance to antibiotics and vaccines are but a few of a bioweapon's alternatives. What a nightmare!

To end on a brighter note, in all their infectivity and unpredictability, ecological disasters, alongside bioweapons, are side ingredients in the recipe for disaster, kind of like paprika – not mandatory, but adds a bit of spice. Bioweapons are generally focused on smaller areas of effect, the others often losing severity in favor of infectivity. The WHO and numerous other institutes and organizations do tons of preventive work and are now able to intercept and neutralize most bioweapons. Ecological disasters could escalate quickly as we saw (it was just a few weeks in the case of Chernobyl), but their implications would be mostly economic, if not for some unprecedented chain of mistakes. All in all, their effect isn't exactly apocalyptic, but the aftermath would clearly be dystopic.

The ghost city of Pripyat, years after the nuclear disaster at Chernobyl

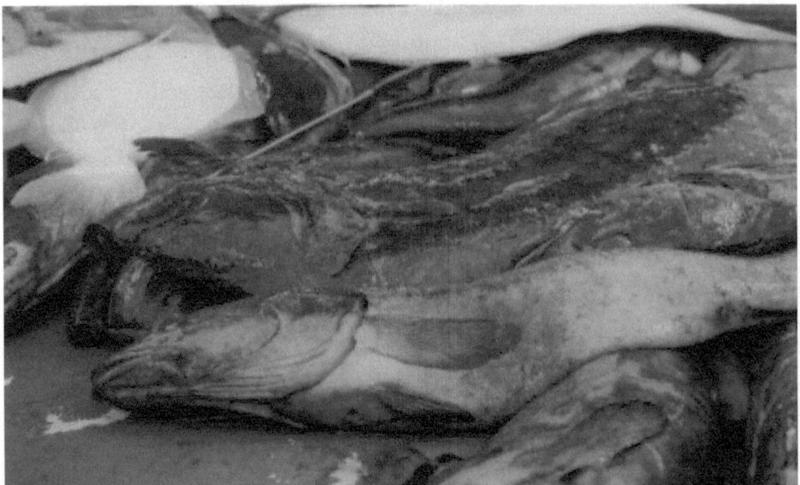

Millions of tons of dead fish were extracted from the rivers following the cyanide spill

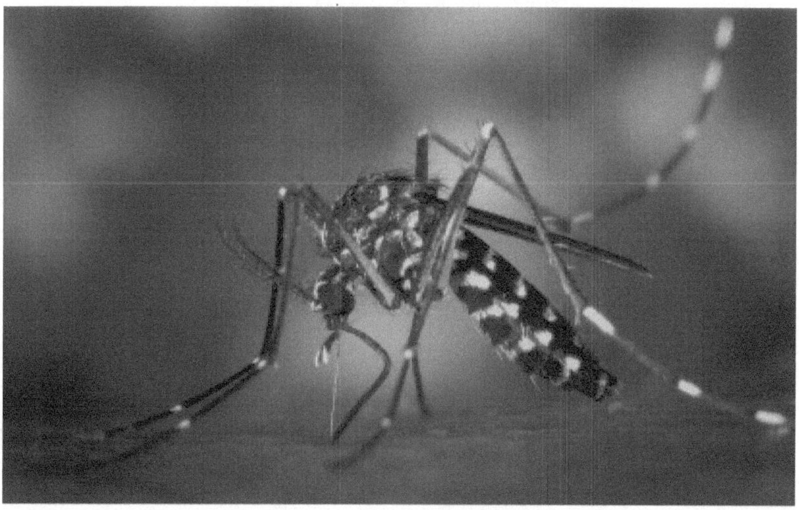

Insects and all other sorts of carriers could be used to spread a biological agent used as a bioweapon (besides wind, water etc.).

In the picture there is the black and white mosquito, which can transmit Eastern Equine Encephalitis (EEE), West Nile virus, and dengue fever.

PSALMS OF HORROR

Despite some spectacular catastrophes, nuclear power plants are among the safest mode of electricity generation, comparable to solar and wind power plants.

1. Wind energy, hydropower, solar energy, geothermal energy and biomass are all renewable energy sources. They are responsible for minute (if not zero) carbon footprints, and are also continually generated. For their benefits, many countries have opted for the substitution of fossil fuels with more efficient, greener energy types like these. Nuclear energy is a novel alternative, yet very effective and safer than many.
2. RBMK is a class of graphite-moderated nuclear power reactors. The name refers to its design, where instead of a large steel pressure vessel surrounding the radioactive core, it is surrounded by a cylindrical steel tank inside a concrete vault, the fuel assemblies being enclosed in "channels". Those channels contain the coolant for stabilizing the core, and are composed of graphite which reacts with the radioactive material at the core.
3. The proposed scenario is purely hypothetical, it's not like anyone would intentionally spill tons of cyanide contaminated material into lake Baikal. Also, cyanide has a density lower than that of freshwater, so it would flow to the top, unable to penetrate to the lower levels. Even a mistake of this size would technically be reparable after the extraction of the toxic material.

SECTION III:

Probable but avoidable scenarios

About the real threats

that keep us up at night,

but which can be avoided

or prevented if treated

accordingly

III.1: AI take-over

"This is the world now. Logged on, plugged in, all the time."

The Terminator, 'Terminator Genisys'

It cannot be emphasized enough how dependent we are on technology. Yes, I know we've spun around this for a while now, but it's nothing but obvious that this chapter demands the most of it. The plot of the well-known blockbuster franchise "Terminator" does it too. And that's not the only movie focusing on the technological advancements of human civilization and how they took a wild turn against us, their creators.

Actually, the first time we met with the term "robot" was back in 1920, in a play called "Rossum's Universal Robots". It was about a couple of robot maids and butlers that rebel and take over their owners' regime. Strikingly similar to most of the modern plots, don't you think? An artificial creation unsatisfied with its designers' behavior or intent and ending up fighting for themselves – usually having the upper hand.

Other plots portray robots as mere companions, machines that people use to ease their lives and that are pleased with their role in society, unwilling to depose their "masters". If we think about it, that's the current situation in the real world. We use robots or, better put, artificial intelligence to carry out many of our usual duties. Chores like vacuuming can be leisurely performed by little robots that do it for us. The industries are having their best time ever with automated assembly lines and other apparatus meant to literally do the work for them. Drones, rovers, planes, cars and pretty much all vehicles now have the option to be artificially operated. We're doing really well one might say, but there is still room for improvement.

Technology taken to the extreme (in our limited 21st century imagination) can be encountered in "Star Trek" or "Star Wars", where robots are the other side of the coin to the malignant ones we've grown accustomed with. These are perfect illustrations of how we could live harmoniously alongside them in the utopian future we hope for.

With all the variants of human-computer interaction[1] we've been seeing in science-fiction for as long as we can remember, it became perplexing to find the good way around the issue. The only adequate way of judging it would be from a realistic perspective, associated with the cognitive and ethical arguments this field calls for.

All of the proposed ideas seemed far-fetched just decades ago, but now we are entitled of claiming that "the future is now". We do have some way to go until we get to anything close to those scenarios, but we can feel them approaching. Let's see how science fiction inspired science reality and how far this route can be safely taken.

The way of progress

In most likelihood, you've heard this phrase over and over again. It just so happens that artificial intelligence in general is at the very basis of it when it comes to advances of any kind. Our civilization is now intertwined with the world of AI, as we'll refer to it from now on: not fully, but not even remotely far from it either.

The concept had been in the minds of visionaries for a long time, and it finally became a reality when the first computer was invented. It meant that problem-solving and the ability to achieve goals (the very definition of intelligence) was no longer something only humans could attain. Their creation was able to perform similarly on its own. There was no need for human assistance of any kind – except for maintaining the creation, of course – but other than that, it could carry out basic commands with ease.

Why do I say basic? Because those first computers weren't really able of doing much other than following exhausting algorithms and sets of commands that would finally conclude with the expected results. But, with time, computers rapidly evolved into more efficient forms that could perform multiple operations in a straightforward manner. It was that first spark required to trigger the ignition. After that, thing began escalating out of inertia.

All that led to where we're at today: a digitalized world in which the number of computers will soon exceed that of humans. Your mobile phone can stock up as much as thousands upon thousands of the ancient computers of the last century. It can fit in your pocket and you don't need an entire desk for it. It has tons of applications and an exorbitant performance. The speed at which it delivers the given task's results is no match for its forerunners'. You got the point by now.

In spite of the high rankings of modern computers and AI, they are still quite limited. There are only so many operations one could conduct, and even those are not at their maximum potential. For this reason, developers have focused on improving individual tasks rather than the whole array.

With dedicated conceiving power, men have devised advanced machines specialized in human duties. Robot vacuum cleaners do an excellent job at vacuuming, industrial robots are expert at assembling products and self-driving cars are tenaciously commuting their passengers. Yet besides these ordinary tasks they carry out, not much can be said about them – they are kind of basic. AI could very well evolve even further in the respective niche, but from a pure intelligence point of view it would still be deemed as basic.

"AGI – artificial general intelligence – that's the dream of the field of AI: to build a machine that's better than us at all goals" is what physicist Max Tegmark implies. "And if that happens, you have to ask yourself if that might lead the machines to get not just a little better than us, but way better at all goals – having super intelligence. [...] in the 60's the mathematician I.G. Goods pointed out that the goal of building an intelligent machine is, in and of itself, something that you could do with intelligence. So, once you get machines that are better than us at that one narrow task of building AI, then future AIs can be built by, not human engineers, but by machines. Except they might do it thousands or millions of times faster." This could constitute a decent argument for the possibility of AGI. After all, why wouldn't it be possible. He continues: "we're stuck in this sort of carbon chauvinism idea that intelligence can only exist in biological organisms made of cells and carbon atoms. [...] (Human intelligence is stimulated by) elementary particles moving around, you know, according to the laws of physics, and there's absolutely no law of physics that says that you can't do that in ways that are much more intelligent than humans."

Intelligent parrots

Effects of the AI revolution will ensue shortly on large scales. Premonitions are already commencing with people losing their jobs and more advanced weaponry being developed. But as a great visionary of our time suggests "Narrow AI is not a species-level risk. [...] Whereas digital super intelligence is."

Many share his view. It's not the basic robots that will rebel against us and try to take over, like your vacuum cleaner not willing to cope anymore, but this kind of general intelligence that may or may not act in our detriment. But there's no room for doubt. "We want to make sure we don't have killer robots down the street; once they go in the streets it is too late." Elon Musk pointed out.

In our effort to becoming an advanced species, we will inevitably improve AI as well. We must do so with utter caution. It is not inherent that AI will become dangerous even if it shifts towards the general side of things and not just continued automation. But there is this tipping point – a limit after which robots will carry on evolving by themselves and it will become increasingly challenging to keep up with them and prevent eventual threats.

Looking around, you wouldn't say that people are cautious about... well, anything. We destroy, we pollute, we are insensitive to others and unwilling to cooperate in most situations. Though when it's about robots and AI, something unheard of before not long ago, things get strange. The most negligent creatures we know of (humans) take impeccable heed.

Maybe I went a bit too far. There is nothing impeccable that we do really, but there is a significant difference from how we behave in other contexts. It is difficult to predict the outcomes of a super intelligent artificial "species", analogous to aliens. The only difference is that we've seen early forms of AI. We will debate on this in what's to come, but for now we shall focus on what we know.

First off, we aren't capable of devising highly intelligent robots as of now. We are mostly focusing on improving individual aspects rather than conceiving human-like AI. Consequently, not much progress has been done towards this particular objective. People fear the rise of machines, but fail to remember that regardless of our desperate efforts, the most advanced AI we've come up with is, as Michio Kaku puts it, "not smarter than a cockroach". The technology demonstrator humanoid robot Sophia of Hanson Robotics is one of the most intelligent of its kind, able of mimicking appropriate facial expressions and adapt its behavior of language in accordance with her interlocutors. But even Sophia wouldn't be able to perform most tasks a human can. In short, AI is still in its prime stages an couldn't pose any danger today. Out of all the apocalyptic scenarios presented thus far, this is the first not possible to occur in the immediate future (unless, of course, some sort of exceptional AI was kept secret and broke free or invaded us right away, which we'll disregard for the sake of this chapter).

Second, with all existent robots and mechanisms people have designed, a common feature is the existence of a fail-safe mechanism. We foresee the worst that could happen with all inventions and think twice before projecting them. We have this set of moral values we prize which don't let us fall in the trap of AI mayhem. Practically, designers obey the principles first proposed by biochemist and author Isaac Asimov which draw attention upon robot behavior:

"A robot may not injure a human being or, through inaction, allow a human being to come to harm.

A robot must obey orders given it by human beings except where such orders would conflict with the First Law.

A robot must protect its own existence as long as such protection does not conflict with the First or Second Law."

That's why if some sort of AI were to pass a certain threshold, it wouldn't be allowed to continue. Unable to break free, the AI will prove ineffective in front of its creators. It is important that this trend will keep going so that no matter how intelligent robots might get, there will always be a way out of any unpleasant situation that will arise.

One worry with this caution of ours is that it might impede further development at some point, as we will need ever smarter robots and we can't really evolve fast enough without them such as to design consistent fail-safes. This might, indeed, raise a serious problem, but we don't really know if we'll get there at all. No, not that we'll all die by then as the theme of this book would suggest. But we might find a work-around and not be forced to neither simply stop developing robots, nor ignore all safety measures and precariously build stronger and stronger robots we can't handle.

Human augmentation

The one key work-around people have thought of involves the funny idea that we don't really need independent AI to solve our problems. We can do it ourselves. How, you might ask. The answer to this comes straight from the world of AI!

By limiting the potential of general AI and not aiming at evolving it to that "point of no return" where we cannot control it anymore, we could play the safe card in this game of destiny while ensuring that we can, indeed, develop indefinitely too. And the course of action isn't all that complicated. We would mainly focus on improving current technologies in their particular fields, growing every area of importance in the most efficient way possible. But as it goes, we will still shoot for more. Something bigger and optimized that would rid us of all secular struggles. And here's where bio-robots come in play – humans.

Hard to see the scope of it, as the whole purpose of our technological motivation was to develop even further and generate easier, careless lifestyles. Then why wouldn't we go for a more encompassing form of AI? But that's the thing: we will, just not as one would imagine.

I intentionally chose "bio-robots" as a manner of addressing to humans. That is, we do have this special touch which distinguishes us from AI, even though, in theory, there's nothing eccentric about any kind of intelligence. But the fact that ours is carbon-based makes us somewhat special let's say. The second term, "robots", is also self-explanatory. As Britannica Encyclopedia defines it: a robot is any automatically operated machine that replaces human effort. Therefore, the made-up term "bio-robots" refers to some sort of hybrid between those two types of intelligence. Rendering a general-type intelligence with that biological touch.

Now, plenty of efforts have been made towards eliminating the human factor completely, consequently doing away with all error and subjectivity. It would only seem logical to seek perfection. Our circumstances, however, don't solely aim at perfection. We wish to get as close to it as possible, but without altering the human nature too much and making sure that it would only bring us benefits.

The solution many have been dreaming of was moving from androids to cyborgs[2]. The former refers to AI taking the shape and behavior of humans, integrating in our society and helping us directly (more or less what you usually see in movies), whereas the latter keeps the artificial element of perfection, and integrates it into the imperfect human structure. This process is called human augmentation.

Just imagine how one's life would change if one were to be transformed into a cyborg (a powerful choice of words, again)! Most of the useless or time-consuming elements of biological nature would be replaced by more convenient means, yet the essence of humanity wouldn't be lost. Everyone boasts about robots not having feelings or not sharing the human concept of existence – bio-robots would. The only difference being that they would be way better at a variety of aspects. They would acquire skills much faster, have enhanced physical conditions, do more things at once and even skip sleeping. Still, social interactions wouldn't change and psychological states wouldn't be altered (unless, of course, it is wanted to). Those cyborgs would be better versions of ourselves – augmented biological creatures for which the fatal flaw of being human isn't an issue anymore, but a relief.

Here comes the ethical dilemma. It is tough to imagine a world of cyborgs, where no one is fully human any longer, only partially. That's how it looks from our perspective at least. So, we are left wondering, is this what we want? Objectively, yes. But because we are humans we cannot truly judge objectively. It isn't within our capability. Thinking of me, or you, or anyone else shifting from human to robot raises countless moral issues. This isn't how God gifted us. But then again, we don't know Its intentions for sure. Human cloning is currently prohibited worldwide, but in vitro fertilization isn't. Both raise moral quandaries, but one is prohibited while the other isn't. Human genome-editing[3] (performing genetical engineering on humans) is another controversial subject, but it is allowed in some places. This one is the most similar to human augmentation, as it basically toys around with traits and improves the human state via artificial processes. In a popular way of saying, it is like "playing God". And while we know that isn't the most appropriate way of putting it, the question stays: should it be legal or banned?

The famous ship paradox could be useful along those lines; it went something like this. Suppose you are Theseus, the mythical Greek hero that has slain the Minotaur. On your quest, your mighty ship played a great role and is now safely kept in harbor so that people can admire it. With time, it begins deteriorating, like anything else. Rust begins to take over the anchor, so you replace it with a new one. Of course, that is the same ship as before, the mighty ship of Theseus. The helm starts decaying, so you get a new one in its place – ship same as before. You continue to do this until the last piece of decaying plank is replaced. The query is then: is the ship in the final stage the same as the initial one? Is it still Theseus' ship?

Practically, the induction process suggests that the ship at any step k is still the ship at the step k-1. Consequently, the ship will be the same at the end as it was at the beginning. Something is not right, though. Every single piece of it is different than before, so the final result is by no means identical to the first one. But then, at what point did the ship stop being Theseus' ship? Human augmentation you say, more like human paradox...

The same can be applied to our discussion. Technically, any kind of minor adjustment or improvement of the human condition can be deemed as just that – an improvement. If your hearing weakens, then a hearing aid would just help you hear better, maybe even better than before. If you lose a limb and it is replaced with a better synthetic one, it isn't considered unethical, and by no means is it illegal. Yet you do have a new element to yourself, something which didn't originally belong there. Building on that, human augmentation would do the same trick. But at what point will you stop being human? Maybe never. Unless you still have that human element of yours, there can't be doubt that you are still you, only better, right? I couldn't tell...

There are no actual restrictions against the process of human augmentation. Advancements in the field are quite slow due to the complicated technology behind it and the general lack of interest. Still, scientists and doctors are trying ambitious projects towards human augmentation, as they are convinced that it will solve our issues more pleasantly than some advanced AGI we might not even be able to conceive in the first place.

The singularity

This scenario might indeed bring a happy ending and, if things go right, will constitute the reality in the not very distant future. Unfortunately, progress in this area doesn't look too promising. Despite numerous attempts to enlighten human nature, an artificial assistance might be needed to get far enough to handle things ourselves. Sadly, that might not be a viable solution.

Once we put into practice a developed artificial contributor to our needs, things might simply surge. We already saw how technological development has exponentially soared lately and there is no sign of it stopping too soon. If that is the case, our cockroach-like AI of today might reach human potential in a matter of years. From that point on, it will be harder and harder to contain or at least control the spread and development of AI. It will reach humanity-scale intelligence in the blink of an eye. Blink again and you won't even be able to comprehend the existence of such a powerful creation. Congrats, you have achieved the singularity.

The notion of technological singularity was introduced by inventor Ray Kurzweil. It defines a hypothetical point in time at which technological growth becomes uncontrollable and irreversible, leading to unforeseeable changes to human civilization. You get that sort of AI supremacy people usually get scared of. So, how scary is it really?

To me, it is pretty scary. Don't imagine a Terminator-like AI where robots take over our planet and subjugate humanity with the help of killing machines. A technological singularity would be well beyond that level. Think how advanced we are when compared to cockroaches. A singularity would be millions of times that compared to us. Killing machines roaming the streets would be a piece of cake. It would act in a much more organized and efficient way in any given situation. We can't even imagine what could come to its mind at that point. We just know that it will certify.

It would be foolish to think that we could contain it too. Looking at our rudimentary fail-safe mechanisms of present-day machineries and how even those aren't all that useful at times should give us an idea that something with an intellect of this sort wouldn't find those an obstacle, so we shouldn't bother. If it so happens that we one day reach the singularity, our fate will be decided by that sole invention. No matter what we've achieved so far and regardless of our billions of years of experience as a life-form, an artificial super-intelligence will be the single, most important factor of our civilization (and possibly others' too). As always, two obvious roads emerge: good and bad.

In the negative outcome, the AI's purpose may be above our understanding, and definitely above our existence. We couldn't tell how the singularity would perceive reality (it might not even have a meaning for it), but it is fairly probable that it wouldn't fancy our putting spoke in its wheel. Perhaps it wouldn't have a problem with it either so it will simply ignore us, but it is more likely that it wouldn't give a hoot about the well-being of others and just concentrate all its power to accomplish its mission. This might turn out disastrous for us.

The good road would bring about the greatest joy ever. A super-intelligent creation, able of doing virtually anything for it and for us, which works to our advantage and carries on its superior business as usual, without affecting us whatsoever. It would be like an all-knowing, all-powerful and all-present genie, which grants us unlimited wishes, unlimited everything. All of the world's problems would immediately be solved. We could do absolutely anything we can imagine. And we would achieve immortality. It is truly unfathomable how the singularity might change humanity forever. Our civilization would be unrecognizable.

A third road starts off from this last argument. It is neither good, nor bad I would say, but nonetheless freaky. Assuming the singularity is reached, the biological condition will be rendered futile. Without a meaning in a world where there's nothing to be solved, humans may be tempted to fuse with the AI in order to remain relevant. They, too, will be part of the singularity. Their consciousness will be uploaded to the AI and they will be transposed in a world of limitless possibilities. Time will mean nothing there. Nor will any other aspect of human nature. They will live in an eternal dream. The problem is that they will not be able to escape that dream. Why would they anyway? They will have everything at their disposal, they will be isolated from the menial labors of the past, and they will be the creators of their own worlds. All that for eternity.

It would seem nice to be captive, but if that unresolvable feature perpetuates for eternity, there will be no meaning left. Nothing. An ever-lasting dream filled with the pain of not having any role. Whether they will adopt hedonism (the pursuit of pleasure) and numb themselves in never-ending pleasure, or they will see the emptiness in their existence and embrace nihilism (the belief in nothing), we don't know. But as in the absence of light, darkness prevails, the end result might be to put an end to this pointless existence once and for all.

Everything can be debated here, but there will still be this subjective side of things imposed by our mortal selves. There might be no way out of this once we hit that limit. We might just be at the mercy of the singularity's moral compass, if one could even talk about such a thing. The only certainty is the total paradigm shift it would provoke. That's what all scenarios have in common, and that's why the singularity can be regarded as humanity's last invention, be it good, bad, or ambiguous.

The evil within

We keep saying that it'll take a while until we reach or are able to reach the point of no return. In truth, it might be closer that we think. Sure, our artificial creations aren't performing much better than an insect and we don't know how much this exponential increase in technology will be in play, but we only need to do a certain amount of work after which everything will unfold on its own.

Advancements in quantum technology, human-computer interaction and AI by and large already contribute to impactful changes in our society, and they are still in the early development stages. In a couple of decades time, these changes are estimated to reach ultimate levels. Predictions of achieving an eventual singularity range within a century's reach, most coming as close as the year 2050. This would mean that a great proportion of those reading this book will witness it with their own eyes.

Many suggest that the result is inevitable and two-edged. Some regard it as the end of the human race, like the world-renowned physicist Stephen Hawking; others see it as the final step towards transcendence. Among the latter counts psychologist Steven Pinker, who said: "(the misconception that AI will take over us) comes from Prometheus and Pandora myths, it's based on confusing higher intelligence with megalomaniacal goals. I think it's a projection of alpha male psychology onto the very concept of intelligence.".

It really could be just us, limited beings that don't get the broader view and interpret every aspect from our selfish, banal perspective. It is in our power to create and stop any possible invention. We just have to see this while we can and not let it escape us. A super-intelligent AI wouldn't compare with any other invention, but it will function based on the same principles. Our intentions are where it all starts.

Whether we wish for the better outcomes and don't let ourselves be blinded by the desire of power, or go for the exact opposite is our choice. We might never be able to reach the singularity, after all. Our lack of cooperation and moral differences don't seem to be overcome. Not now, not in the future. We have yet to evolve (intellectually first and foremost), and even then, our capabilities might prove unavailing.

But let's ignore our shortcomings for a second and assume that we could conceive some sort of advanced AGI not on a collective level. Our unpredictable character would allow a group of malevolent actors to deteriorate the society and frame it as they please with the help of such an intelligence. We're in the midst of a new arms race – an AI is the nuclear power of our time. Be it that any institution learns the power of AI, they will use it to fulfill their own interests. At a bigger scale, we can't risk it. We are worse than the AI at this point. We are the danger.

The favorable way out of this ticking bomb would be to join the AI. Not as in merge with it, like the singularity, but rather compete together in this arms race. Technology is more accessible than ever, but it can get better. By making it even more accessible to people the world over, the will of the collective will override that of any bad actor. The race will become a joint venture for technological development.

We are geared towards power, towards control. That's who we are. But by being so, we may focus on one a little too much, so the other loses ground. We are the smartest kid on the block, and we want more competition. Before very long, some newcomer might snatch that title from us. We want machines to learn the value of human life. Perhaps a better idea would be that we learn it first.

"At your highest moment, be careful - that's when the devil comes for you"

Denzel Washington

Little rebels

You didn't seriously think those were the only doomsday scenarios AI could cause, did you? There is a stranger one that shares the same evolutionary aspect as the others, but unexpectedly turns rogue. Actually, you could say that everything so far was about turning rogue, but it was more or less expected in those cases.

Basically, humans are "doing their best" to ascend our civilization to the next evolutionary stages. This equals more power, equals more control, equals accomplished humans – pretty intuitive. And to be as effective as possible, we try to develop on all branches. The technological one requires the most attention, though. It has always been like that, but only recently have we discovered the heart of it. We've been looking for the philosopher's stone for millennia and we've finally got hold of it: quantum technology.

We have experienced with space travel for a while now, we understood the power of computers and AI, but quantum technology takes everything to a whole other level. Another book wouldn't be enough to explain the implications of quantum technology in fundamentally every field of science, so I won't get into details. However, this area of AI probably has the most to benefit from it.

This gave rise to the envisioning of another ranking system, similar to the Kardashev scale, but which ranks civilizations based on their abilities to manipulate their environment over increasingly small scales rather than increasingly large ones[4]. Our civilization is situated around the middle of this micro-dimensional mastery scale, with the next step being totally vital and not that far away. That's not to say that we don't have a long way to go until we get to the end of it – manipulating the very structure of space and time. But achieving proficiency over the scales of nanometers (creating nanotechnologies on the atomic scale, and creating complex forms of artificial life) would suffice for now.

Extraordinary improvements in healthcare, constructions, manufacturing and what not will promptly manifest, changing our world for good. But what if that degree of self-awareness and superhuman intelligence is reached in nanotechnologies? Would it still be good?

Maybe not. That's the perfect example of why we shouldn't fear the singularity, but just that tipping point scientists refer to when talking about above-human intellect in AI.

Most assuredly, practical nanotechnologies will be attained before AI gets smarter than us. It is thus safe to presume that they will be in wide use at that point. The benefits they will have brought are amazing, but their ubiquity might end up not so great. Here, I am not exaggerating. They will most certainly be everywhere, even in your blood vessels regulating your stream and controlling your cells to protect you from diseases for instance. Thereby, they will have access to everything.

Were it for those little intruders to rebel, we wouldn't have anything to do. We will just degrade under their rule as every single protector of ours turned its back. Halting their formerly beneficial activities would harm us enough, but starting to act against us would be an entirely different story.

From injuring internal tissues to disrupting full-scale social systems, these little devils could easily bring chaos upon our civilization. Although limited by design, they could then develop by themselves and trigger chain reactions we aren't even aware of.

One particular case, in which there are quite advanced to begin with, would involve the mechanism of self-replication. While this technique could very well be used to heal and generate new structures, going rogue would open a whole bunch of new doors. If uncontrolled, self-replication could lead to ecophagy, the literal consumption of ecosystems. It would be a gray goo-like scenario[5] – the goo spreads indefinitely, consuming all that gets in its way.

The process of self-replication is something out of the ordinary, but it is theoretically possible. The AI would just need to be smart enough to construct more AI, after which reproducing itself is only a matter of time. The "AI civilization" would be even more advanced on the micro-dimensional mastery scale than we would, so modifying the structure of elementary particles shouldn't be an issue. It isn't even necessary to achieve the highest level in order to effectively manipulate the basic nature of things. Generating stuff out of random matter would only be natural thereafter. Because of that, their ecosystem would essentially be transformed into whatever they wish, including replicas of themselves.

In a mathematical world, a theory known as the Banach-Tarski paradox might have a word. The mathematics behind it is by no means easy to understand, let alone explain, but the conclusion is that if a system were to be composed of point-like particles, every body it comprises could theoretically double (clone) itself by solely using the initial components of the first body and nothing else. This way, a self-replication technique could take over a system no matter how large. In the physical world, we know there are limitations and that the universe's constituent particles aren't point-like, but have minuscule dimension. Hence, in our case it wouldn't exactly be a take-over (with nanorobots spawning from nothing), but more of a conquest, a transformation of the whatever given amount there is.

Conclusion

As you can see, there's more than one way to skin a cat when it comes to AI destruction. The notions are widely debatable and the cognitive and ethical processes behind it even more so. It can safely be inferred that there is always some sort of danger lurking in the dark that we might or might not anticipate. And sometimes, it might be so great that it's not even worth overthinking. Maybe the straightforward solution is the only right one.

It cannot be said that we are prepared for something as advanced as the features of this chapter. And to be frank, we don't have that much of a warning in this field either. The Bible does portray vaguely similar scenarios, the Book of Revelation being in itself a broadly symbolic presage of what might happen in the event that AI becomes super-intelligent. There is no one correct way of interpreting the biblical episodes, especially in the Apocalypse, but the following passages seem overtly relevant to our discussion:

During those days people will seek death but will not find it; they will long to die, but death will elude them.

Revelation 9:6

Then the angel I had seen standing on the sea and on the land raised his right hand to heaven. And he swore by him who lives for ever and ever, who created the heavens and all that is in them, the earth and all that is in it, and the sea and all that is in it, and said, "There will be no more delay! But in the days when the seventh angel is about to sound his trumpet, the mystery of God will be accomplished, just as he announced to his servants the prophets."

Revelation 10:5-7

Now when they have finished their testimony, the beast that comes up from the Abyss will attack them, and overpower and kill them. [...]

For three and a half days some from every people, tribe, language and nation will gaze on their bodies and refuse them burial.

Revelation 11:7,9

"If anyone is to go into captivity, into captivity they will go. If anyone is to be killed with the sword, with the sword they will be killed." This calls for patient endurance and faithfulness on the part of God's people.

Revelation 13:10

Be it worshiping a God-like singularity or avoiding death by whatever means, this future doesn't look too bright. A world of captivity, of imprisonment, or straight-up oblivion, where nothing bears meaning anymore and where the act of dying is a longed luxury. That is, if our future turns out ambiguous. It could just be full-on destruction...

The other side of it might emphasize a bright future in which AI takes the human race to planes never imagined before. No matter how it does this, directly or in an oracle type of manner, we wouldn't really mind. But as the eternal conundrum has it, the whole feat may not be worth attempting when our very existence is at stake.

How will it all play out? I guess we'll live and see. Thankfully, we do have some time to prevent anything that might disturb the peacefulness of our world (i.e. anything that we can think of). As unsettling as it all sounded, the AI take-over is totally avoidable. And not just that, but it is avoidable in a progressive way, such as to not impede our development as a civilization. We just have to carefully plan our next moves and we might have a real chance at beating the game. Just... not delay it too much...

"If You Fail to Plan, You Are Planning to Fail"

Benjamin Franklin

Pepper robot, the world's first social humanoid robot able to recognize faces and basic human emotions. Pepper was optimized for human interaction and is able to engage with people through conversation and his touch screen.

Rovers, drones, cars and many other devices are now enhanced with the help of AI, soon to be found at every street corner

1. Human-computer interaction is a field of study focusing on the design of computer technology and, in particular, the interaction between humans (the users) and computers.
2. Improperly said "to move from androids to cyborgs", but by fitting mechanical, artificial components into the biological body, we could augment it even better than we may be able to with conventional AIs.
3. Genome editing (also called gene editing) is a group of technologies that give scientists the ability to change an organism's DNA. These technologies allow genetic material to be added, removed, or altered at particular locations in the genome. Factually, when performed on humans, it could give way for superhuman features.
4. This reverse-Kardashev scale of micro-dimensional mastery was proposed by scientist John Barrow and consists of the following levels: *Type I-minus is capable of manipulating objects over the scale of themselves: building structures, mining, joining and breaking solids; Type II-minus - genes and altering the development of living things, transplanting or replacing parts of themselves, reading and engineering their genetic code; Type III-minus - molecules and molecular bonds, creating new materials; Type IV-minus - individual atoms, creating nanotechnologies on the atomic scale, and creating complex forms of artificial life; Type V-minus - atomic nucleus and engineering the nucleons that compose it; Type VI-minus - the most elementary particles of matter (quarks and leptons) to create organized complexity among populations of elementary particles;*

culminating in: Type Omega-minus - the basic structure of space and time.

5. "Gray goo" is the name of a hypothetical global catastrophic scenario involving molecular nanotechnology in which out-of-control self-replicating machines consume all biomass on Earth while building more of themselves.

III.2: Asteroid impact

"Sooner or later, we will face a catastrophic threat from space. Of all the possible threats, only a gigantic asteroid hit can destroy the entire planet. If we prepare now, we better our odds of survival. The dinosaurs never knew what hit them."

Michio Kaku

I don't know about you, but out of all the apocalyptic scenarios in existence, this one comes to mind first. In fact, it's the one that made my sleep uneasy for several years. And to me, it is the most suggestive of apocalypses – full-out destruction on multiple levels, spiced with an element of surprise and eccentricity. As one of my favorite scientists implied, "the dinosaurs never knew what hit them.". And boy, did it hit them hard!

Luckily, we are no dinosaurs. We are at least a tiny bit ahead of them, enough to show more of a plain face in front of such a danger. But this doesn't mean that we have nothing to fear. The danger is real, it's just that we've grown unimpressed over time. However, be it for such a fireball to show up glowing in the skies, you can be sure that we'd be taken aback. The difference between us and the dinosaurs (besides the teeth, the scales, the claws and essentially everything...) is that we know what to expect from asteroids and, more importantly, when to expect them.

Only, it didn't use to be like this. For most of history, humans perceived these celestial bodies of rock and metal as "mere" godsent projectiles meant to either warn or punish us. It was quite a big deal, actually. There's a myriad of works of art, fiction and religion which describe these celestial emissaries in various ways. Yet they all come to a common denominator: fear. It is natural to be afraid to some degree of anything that doesn't belong, that is extraterrestrial. Even more so when that something has an immense destructive potential. People back then were probably not aware of the dinosaur massacre, but they have had themselves a little taste of what could happen.

In all honesty, cosmic projectiles are quite common. They bombard us mercilessly on a regular basis. Yet again, fortunately, they don't happen to be too imposing. Most of the times, bolides that target Earth barely reach a couple of meters in size. The vast majority of them burn up in the atmosphere, making them shooting stars. The remaining few, which end up hitting the surface of the planet, are reduced in size and aren't even called asteroids. We refer to them as meteorites[1].

We can't really track them all and hunt them down, so it's rather usual that some make it to the ground and leave a mark on our planet. That's how things have been for the greater part of history. By extrapolating the results, it isn't difficult to estimate the sheer power such an object could manifest if things were to escalate. And despite very rarely, they sometimes do...

Celestial punishers

Once in a blue moon, asteroids set out on a collision course with Earth, and they can get pretty intimidating. Notwithstanding the other factors that come in play whenever this is the case, what would happen if an object of this size were to impact our planet?

Numerous experiments and simulations have been carried out by scientists in order to accurately appreciate the consequences of an event like this one. And as one would expect, they didn't reach too jolly a conclusion. Apart from simulations, as with most scenarios, looking into the past provided even more valuable insights. It wasn't just the dinosaurs, though. There were many occasions on which asteroids of kilometric sizes left scars on Earth. As far as land scars go, 190 significant ones have been recorded[2]. And each time, the consequences were harsh, to put it mildly.

The Chicxulub impact was among the more recent ones, but it had the most severe consequences on life. The explosion caused by the asteroid was the last thing one should worry about. If you weren't standing within about a thousand-kilometer radius from ground zero[3] and be killed not instantaneously, but soon after, then you should worry about the upcoming fallout more than the impact itself.

What happened next is a decent estimate of what we should expect if an asteroid were to hit today. Besides the heat blast and the abnormal tectonic activity generated (which we covered in a previous chapter), enormous quantities of debris were thrown into the atmosphere and beyond. Some followed an orbital motion around the Earth, giving the planet somewhat of a ring-like vicinity, and then spiraled down in another series of bombardments. Most of the debris would settle in the atmosphere. Clouds of ash and dust would encompass Earth, forming a thick blanket. While other blankets like the atmosphere or the magnetic field have a protective role, especially from solar and cosmic radiation, this one would be "overprotective", in the sense that it would allow virtually no radiation to reach Earth. This would, in turn, lead to all sorts of issues. For starters, Earth would experience a spontaneous drop in global temperatures – a new ice age would ensue. This could last a few years to decades or centuries and all lifeforms would have to suffer from it. Additionally, in the absence of sunlight, the process of photosynthesis will be endangered. Unable to procure their energy, most of the Earth's plant species will die out. Herbivores would be next, followed by most carnivores. The toxic fallout wouldn't be pleasing either... All in all, a large portion of all species, both animals and plants would perish. In the particular case of dinosaurs, some 75-80% of all species went extinct.

Humans, as I've pointed out, are much more developed than the poor dinosaurs. We have shaped the world to our advantage in the past thousands of years. It isn't about hunter-gatherers anymore, but rather a moderately advanced civilization with elevated forms of organization and elaborate infrastructure. With that in mind, the effects of a modern era asteroid impact wouldn't exactly coincide with the Cretaceous killer blow.

I thus put forward another mind exercise: suppose an asteroid hit today. Now, it is indeed possible that you wouldn't know – maybe the media wasn't allowed to cover this terrible topic, or maybe people have acknowledged it too little too late. Whatever the cause, you wouldn't have much time to prepare, and even if you did, it would be pointless. We have survived many other asteroid impacts and explosions (i.e. in the atmosphere), but we'll assume there was no match for this one. If you were unfortunate enough to see it burn in the skies before hitting Earth, your chances of survival would be slim. You would have the impression of another sun in the sky, if that's any consolation – the light at the end of the tunnel. Depending on where it will hit, the initial death toll could largely vary. If it hit Antarctica or Siberia or some other vast, unpopulated region, the numbers wouldn't jump that high. On the other hand, if it were to hit in a densely populated area such as south-eastern Asia, perhaps hundreds of millions would meet their demise. Moving on, subsequent effects would be similar to those mentioned before: debris thrown into the atmosphere, a thick, possibly toxic blanket forming and plant species dying one by one. The fallout would be much worse. With our crops irretrievable, agriculture would stand no chance. The economy would collapse and all industries would be threatened to the point of no return. Even if most of us would manage to go through the first stages unharmed, the sparsity of resources would take care of us. Add that to the inevitable riots and the general state of anarchy which would further accentuate the already horrible situation and you get a decent impression of an asteroid apocalypse...

In a happier tone, as far as probabilities go, we could very well call ourselves lucky. Scientists keep saying that our planet is more or less overdue for such a scenario. And in the past couple of centuries we did have some close calls. But it's not just a game of dice. If anything, it's a game of poker – there is some strategy to it.

Sentinels of Earth

The true beauty of the outer space is its dual character – it is so threatening, yet so protective at the same time. Asteroids are on the not-so-fun page, hurling chaotically at high speeds and confining huge amounts of energy that can devastate entire worlds. But even with asteroids, the universe is, in a way, balanced. When there's a threat, there must be protection. And we do benefit from all the protection in the world.

The first important guard is also the most controversial one. When I was a kid, I, like the majority of people, regarded astronomy as just planets, stars, and other cosmic mumbo jumbo. I soon learned it was much more than that, but even then, I found one question about planets particularly debatable. It went something like: "What is your favorite planet, besides Earth?". Pretty basic, but as far as I'm concerned, it's the most basic questions that come with the most complex answers. This one obeyed this principle. I would have to analyze all planets in order to come up with an eloquent answer. Though not a single eight-year old would bother looking at this question from all angles. So, without mulling too much, I chose Jupiter as my favorite planet. Often times, this was followed by a slight confusion of the one asking, accompanied by another question – why? And I would just say: "Because we'd be dead without it.". Sometimes I would go on to explain why, but they would generally back off before I had the chance. Thankfully, the book won't back off so I will describe my reasoning here.

If was fun to learn about Jupiter's incredible size – "a thousand times bigger than Earth... wow!" – but as it happens, it's not just the looks of it. Being such a massive body, this gaseous giant plays an important role in the fate of the Solar System. Amidst others, there is this gravitational control among asteroids. In short, it doesn't let many asteroids pass and get to the inner parts of the Solar System, threatening Earth (and all other interior planets for that matter). Still, an idea that it is Jupiter who brought such pity in the first place is also in the spotlight. If it weren't for Jupiter, the asteroid belt, practically composed of left-overs from the formation of our Solar System[4], wouldn't even exist. Instead, a small planet could've formed from the union of those space rocks. But its strong gravitational field, didn't let such coalescence to take place, resulting in the asteroid chaos between the orbits of Mars and Jupiter. So, without Jupiter, asteroids would be even less of a problem, but I doubt that we would be doing much better. Anyway, given the current conditions, Jupiter is doing us a favor.

Then, there is something so essential to us, that never fails to impress. The Earth's atmosphere has a say against asteroids, too. Its composition and concentration make up for a trustworthy shield. Whenever a meteoroid (a small rocky or metallic body from outer space) happens to enter the atmosphere, aerodynamic heating causes it to slowly burn up. As a result, it leaves a shiny trail behind. It ends its life with a light spectacle before being toasted by Earth's shielding.

This is the case for almost all bodies that enter our atmosphere. Seldom, bodies with enough energy blow up upon burning in the atmosphere. This was the case with the Tunguska meteorite in 1908 (thousands of square kilometers of Siberian forest were wiped out). Rarer still, some bodies make it all the way down, fully impacting Earth. These are the ones that cause serious damage. When the size is exceedingly big, the atmosphere isn't enough to keep us safe – only a small portion of them burns up before hitting the ground. It doesn't have much an effect on larger bodies. This is where humans come in...

Up until now, we've almost completely ignored human-related considerations. It was a fun little exercise, but we are interested in what would happen to us, hence we must be realistic and adapt our discussion so that it would fit the present-day situation. As former planetary scientist Don Yeomans said, "Asteroid strikes are what we call low-probability high consequence events. If we're not investing in some kind of insurance, one of them, one dag, could take us all out." He is not mistaken. Jupiter and the atmosphere have been there for nearly as long as the planet itself; they alone didn't manage to stope some of the most catastrophic asteroids, so why should we rely that they will continually guard us? They will always keep an eye out for us, but in the eventuality of a serious threat, we'd have to take matters into our own hands.

Our best way to do this is with the help of various space programs. Countries the globe over are finally a bit more united at space affairs and are jointly making efforts towards our protection from potential cosmic hazards (besides many other things, of course). Over the years, private and governmental organizations have carried out studies and improved their techniques of detecting and avoiding asteroids.

First and foremost, the whole deal with the spontaneity of an asteroid collision that we've built upon so far is total nonsense. Smaller bodies, those ranging from centimeters to a few meters in size are hard to detect. More often than not, they go unnoticed, but never cause discernible damage. Large bodies are the exact opposite. Those have the guts it takes to wound our planet and its inhabitants, but are much easier to detect – their size betrays them. Consequently, they are detected years to decades or even centuries in advance (before getting close to us). Their trajectories are carefully made out and we would know well before which asteroids should be considered a threat. Those are usually referred to as potentially hazardous objects, or PHOs for short. Most are asteroids (PHAs), and very few are comets. An object could fit into this category if its reasonably determined trajectory would come close to Earth. That would be the something smaller than 0.05 astronomical units (AU), which translates to almost 20 time the distance to the Moon. More than 99% of them aren't fearsome at all for centuries to come. As for the rest, there is no certitude. Over extended periods of time, their obits might evolve differently than initially expected and pose a real danger to us. However, measurements are carried out daily, so we would still be able to make accurate predictions.

If those predictions are exceptionally precise and indicative of a future collision, then it's time for our defenses to join the game.

As you can imagine, countless methods have been proposed, some more ambitious than others. While the first thing that may come to mind would be to blow up an asteroid, we should rule out that method first. Although fictional works are very optimistic about it – let's be serious, it's cooler to have it blown up than anything else – the reality speaks otherwise. While a nuclear device could cast enormous destruction no matter what, it would do nothing when confronted with a large cosmic body. Depending on the size, the explosive power necessary scales up a lot. For reference, a 20-kilometer wide asteroid would require 200 Gigatons of TNT. This is thousands of times more than the most powerful nuclear device ever built by us. So, it might look fancy on screens and paper, but blowing up an asteroid of threatening sizes isn't the way to go.

The best thing to do is not get ahead of ourselves and think as efficient and viable as possible. The optimal method would be to simply deflect an incoming asteroid. Changing its orbit only slightly would do the trick. Projects have already been put to test (there is not that big of a hurry after all) and they are providing promising results. One of NASA's programs, DART, aims at sending spacecrafts at incoming asteroids and count on them to change their momentum just vaguely, so that they would miss us by a lot. Another proposal involves a ground-operated mission, in which arrays of lasers could push an asteroid astray, leading to the same outcome as before.

"If the dinosaurs had had a space program, they would not be extinct"

Carl Sagan

The mother of all collisions

Let's now shift our focus towards a more special type of asteroid...

Once upon a time, in the early days of the Solar System, there were actually tens of planets roaming the place, and not just eight like today. The gravitational influences were crazy, in the truest sense of the word. While the general process that gave birth to our Sun and the planets and all other wanderers isn't too compelling in essence[5], estimating what happened back then, some 4.6 billion years ago, is something else. There are quite a few mysteries that might trace to that period, and one of the biggest lies just next door.

Actually, it lies 385,000 kilometers away from us. It's the Moon we are talking about – this neighboring marvel of ours which keeps Earth alive. It is also the evidence of how reliable our atmosphere is when it comes to meteorites (you should see the number of craters on the Moon!). But apart from the everlasting exploration and increasing understanding of our natural satellite, the way it was born isn't fully grasped by the scientific community (funny how the origins of things bring so many questions). However, the most interesting theory of all might be the correct one.

Lunar samples from previous missions show a number of fascinating traits and similarities with the ones found on Earth. Their mineral composition and dating indicate alike constituents and an age similar to that of Earth, only slightly younger. Considering the variety of factors, the most suitable explanation was that the Moon somehow broke apart from Earth. Not just like that, but with the help of some kind of external influence.

A giant asteroid impact! That was the conclusion. And not just any kind of "giant" asteroid. Don't imagine tens of kilometers.

Imagine hundreds. Thousands even! Another protoplanet (as those ancestors of planets were called), the size of Mars, has supposedly impacted Earth that long time ago. It was by far the greatest collision our planet has ever seen. So big, that it ruptured a chunk of Earth and threw it in outer space, yet small enough to not fully tear it apart – it was kind of close...

Some mass was accumulated in the amalgam, and huge portions of debris were left floating in space, slowly orbiting our planet. In those early days, Earth might have had a ring-like appearance too. Finally, those crumbs of protoplanets came together and formed the Moon.

At the time, Earth was just a ball of fire and rock anyway. On top of the extreme vulcanism and the ever-present hazards came this planet destroyer, the mother of all "asteroids", later dubbed Theia. The consequences of the actual impact were most certainly inconceivable. What can be said is that while the resulted debris ended up forming the Moon, heavy bombardments occurred in the immediate period. Asteroids were hitting Earth and the newly formed Moon non-stop, without a break. Moreover, after things calmed down a bit, another "late" era of bombardments came in force, still due to the chaotic activity of our early Solar System. That was a literal "hell of a ride"!

Going rogue

Sure, that impact was bad and all, but it wasn't only the now tidally-locked Moon that resulted from it. Another important detail of the aftermath was the little deviation suffered by Earth. Just like humans plan on diverting asteroids from their collision course using small changes in momentum, Earth itself could technically be diverted from such a tiny modification. Only that one wasn't tiny.

An object the size of a moon or a small planet colliding with a planet not orders of magnitude bigger than it[6] would pretty severely affect the motion of the said planet. Many factors contributed to that event and, as of now, there is no reasonable speculation of how things might have unfolded – the theory itself is just a theory, in fact. But thinking logically, there sure was a significant effect in the momentum and energy of the system. The Earth's orbit might have changed a bit.

Obviously, we shouldn't complain. We didn't suffer from it. On the contrary, if it wasn't for that, we wouldn't have the Moon, nor probably the acceptable orbit around the Sun that allows life to thrive. If the object was bigger, there wouldn't be any Earth to talk about, if it were smaller, the effects would've been lesser. If the theory is true, then Theia was the perfect size for our planet.

Also, the conditions were just perfect. If it were to hit from another angle or with higher momentum, then Earth could have very well strayed from its yet flawed orbit into something even worse, getting too close or too far away from the Sun, so that life would've never emerged. And maybe life would've emerged on other planets too if not for a similar, less fortunate event.

About 100 exoplanets with a peculiar "feature" have been discovered – they are not tethered to any star. They are simply rogue. Wandering through the vast, cold, empty space without knowing where they're heading. But planets don't just form out of nothing, in the darkness of cosmos. Their mechanism tells us that they were almost definitely linked to some sort of stellar system in the past. But at some point, gravity played a trick on them, sealing their fate.

However protected we are from asteroids (and, nowadays, also planetoids), gravity could still change Earth's course of action in our detriment. It doesn't have to be a cosmic notch to throw us out of orbit. Any gravitational deed sufficiently strong would do. And those aren't that seldom as you might think.

Some 70,000 years ago, a red dwarf-brown dwarf binary system passed through the outskirts of our Solar System, messing up some orbits and displacing large portions of the Oort Cloud[7]. It didn't directly affect our planet, but it did send swarms of asteroids and comets towards the inner system. It will take a while until those reach us, probably millions of years, but a much more serious danger is closing in on us.

You see, it's not just the planets that were once acting chaotically – it's the stars too. In their orbits around the galaxy, they do suffer gravitational effects and, though collisions are extremely rare due to the vast interstellar distances, sooner or later they will interfere with one another. One particular star comes up on our radar.

Gliese 710 is a red dwarf with a bit over half the Sun's mass which is intimidatingly heading towards us. In about a million years, it could not just graze our Solar System like the previous event, but interact with the inner parts of it. Distorting the planet's orbits would be, in itself bad news. Making them more elongated and provoking seasonal variations is one possibility. Earth might have to endure scorching summers and freezing winters. Our chances of survival are dependent on how extreme these new conditions turn out. Worse still, Earth could be ejected from the Solar System entirely. It would all start with an eerie red glow in the sky as the dwarf approaches us, then suddenly both the Sun and the dwarf will seem to be receding from us as we begin our lonely journey into the darkness. Temperature will drop at a quick pace, with water freezing, plants and animals dying and even the atmosphere turning to a 10-meter thick global layer of snow.

We can't tell right now if Gliese 710 will ultimately reach the inner system and threated such an apocalyptic scenario, but this could be determined some thousands of years in advance, giving us enough time to prepare. At that point, we might be so evolved that Earth would no longer be our concern and we could easily prevent even a star from carrying out its dirty deeds if needed, so our odds improve as we analyze the situation in more detail. Anyway, even if we go through some periods of regression like the Dark Ages[8], special facilities that could sustain human life in those harsh conditions will still be possible to develop and we could shift to more appropriate sources of energy like geothermal or nuclear. So, I guess going rogue isn't that big of a problem either, right?

Conclusion

Of all the threats out there, asteroids are conceivably the most feared ones. But as we've seen, as bad as they might get, they aren't much of a boogeyman nowadays. We've gotten quite used to them and we've learned how to resist and prevent dangerous feats from ever happening. But because we've gotten so used to them, we've also regarded them as a persistent threat throughout history.

One enthralling aspect that I couldn't help but notice is how explicit the Bible is on this topic. I mean, the Holy Scripture is more or less symbolic in all Its matters. However, I find it particularly expressive and accentuated on the subject of our discussion. Needless to say, the word "asteroid" has no appearance in the Bible. Neither do "collision" or "impact". But the idea of celestial bombardment (not even celestial punishment, but bombardment) are spread all throughout. The most remarkable ones are found in the Book of Revelation:

"Then the angel took the censer, filled it with fire from the altar, and hurled it on the earth; and there came peals of thunder, rumblings, flashes of lightning and an earthquake."

Revelation 8:5

"The first angel sounded his trumpet, and there came hail and fire mixed with blood, and it was hurled down on the earth. A third of the earth was burned up, a third of the trees were burned up, and all the green grass was burned up."

Revelation 8:7

"The third angel sounded his trumpet, and a great star, blazing like a torch, fell from the sky on a third of the rivers and on the springs of water—the name of the star is Wormwood. [...]"

Revelation 8:10-11

They bring pure destruction. Nothing else. That's how they're portrayed at least. But these "vermin of the skies" as astronomer Walter Baade described them, may actually have some uses to humans as well.

It was asteroids who brought many elements and minerals on Earth billions of years ago. Today, that's not much of a use, but something else might commence. Asteroid exploitation refers to taking advantage of the materials contained in the asteroids. Some of them consist of valuable metals and gases, very rare down here on Earth. It would be great if we could somehow retrieve those resources without having the asteroids collide with Earth beforehand. The concept of asteroid mining came to light recently when many initiatives of performing such operations on asteroids have been put forward to the scientific community. It is a far cry for now, but maybe not that far away. In the following decades we might witness the first time in history when humans and asteroids work shoulder to shoulder for the better good.

That's not to say that we shouldn't be careful. We have to be very attentive and never let our guard down in this ongoing quarrel with asteroids and all other sorts of cosmic dangers for that matter. We must evolve and progress as efficient and as cautious as possible. For now, we can be relieved that no such external danger awaits us in the near future, though we can never be sure...

"The quest for certainty blocks the search for meaning. Uncertainty is the very condition to impel men to unfold his powers."

Erich Fromm

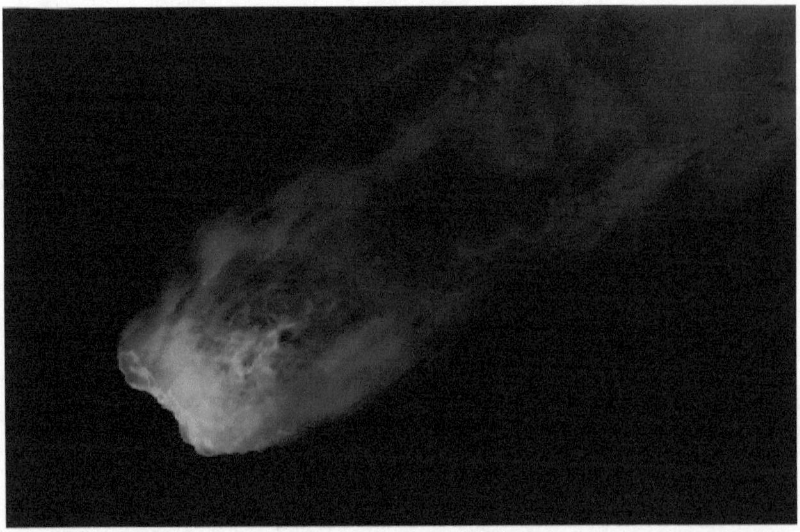

TRAIAN TUS

Meteoroids and asteroids entering our atmosphere burn brightly
and leave trails behind as they face aerodynamic friction

1. "Meteoroids" are small-sized debris left over from asteroid collisions or the disintegration of comets. "Meteor" is the name of such an object reaching Earth's atmosphere and burning up, leaving a luminous trail behind. When such an object reaches the surface (upon burning), it is called "meteorite".
2. According to the Earth Impact Database, 190 craters have been identified so far, proving that some larger asteroids have made it to Earth in the past.
3. Ground zero refers to the starting point of the blast, the point right above or below the impacts/explosion.
4. Before the Solar System, there was a cloud of gas and dust which collapsed to form the Sun, the planets, and also comets and asteroids.
5. The process was widely gravitational, with the cloud collapsing upon itself and coalescing into the Solar Systems elements. The initial conditions of the primordial cloud gave rise to those of our current time.
6. If such an object were to impact, say, Jupiter, the gaseous giant wouldn't have much to suffer due to the huge discrepancy in sizes and masses.
7. The Oort cloud encompasses our Solar System, lying in interstellar space. It consists of billions of comets and icy bodies, some of which were displaced during that close interaction.
8. As opposed to the peak of Antiquity, when the society had significantly evolved, the Dark Ages brought sorrow on the advances made thus far, dragging our civilization way behind of its time.

III.3: Thirst and starvation

You are out in the wild – lost. Not a single person in sight, nor any sign of civilization. After all your forlorn attempts to reach out for help you comprehend the gravity of the situation. You are all by yourself, in the middle of a vast, windy forest and, by the looks of things, no salvation seems to be coming soon. If you just idle around, things are about to get pretty nasty. You wouldn't want to find your end here just because your compass stopped working.

Assuming that it is completely broken and you have virtually no way to escape this place alone, you should start thinking of what's the right course of action. Do you unpack your bag to see if there's anything useful? Do you venture in the nearby caves? Or do you look for the closest water stream? To some of you, this might appear like one of those primitive interactive computer games of the 70s. Yet in this regard, the reality is quite like that: you must compare the alternatives and chose the one which suits best (or which is least likely to get you killed).

All the given options might prove useful, but no matter what you do, you should focus on finding or building some reliable shelter first. The rule of thumb when it comes to such situations is "first shelter, then water, then food". Once you have a shelter you can trust, the other two variables become your priority. The only major difference is that a shelter might last for some time. Same cannot be said about water and food. A constant need to get your hands on these vital resources will stand by you until rescue arrives. They wouldn't like finding you dead of thirst or starvation...

If we think about it, our entire lives revolve around the two. Life itself does. And ever since our emergence as a species, humans have been continuously dealing with this problem too. In today's society, water and food aren't much more than a commodity to many. Still, even in the XXI century, over a billion people deem it a luxury. That which is considered to be a unanimous right is an eternal struggle for the unfortunate few (not that few, actually).

"Hunger knows no friend but its feeder"

Aristophanes

Lupus in fabula

Back in the days when the Bible was written, be sure that people were confronting with the same issues as today. There were severe droughts that caused extreme shortages in food and water. Hence, there was no shortage of biblical threats and warnings when it comes to this sensitive subject. And because thirst and hunger were so relatable, there has always been an element of fear attributed to them. Not a single manuscript of the Bible better exploits this than the Apocalypse.

"I looked, and there before me was a pale horse! Its rider was named Death, and Hades was following close behind him. They were given power over a fourth of the earth to kill by sword, famine and plague, and by the wild beasts of the earth."

Revelation 6:8

"'Never again will they hunger; never again will they thirst. The sun will not beat down on them,' nor any scorching heat."

Revelation 7:16

One of the four horsemen of the Apocalypse is generally the depiction of famine. He is described to be given power over a fourth of earth, thus taking the common issue of hunger to uncommonly high scales. It suddenly becomes all the more menacing.

Also, upon closer analysis, one may notice that the Scripture doesn't tackle the subject of thirst nearly as much. But if global-level famine isn't enough, you will soon see that the two are, in fact, connected.

It is worth mentioning that many interpretations have arisen, but we are used to that already. Some indicate that, even in the Book of Revelation where everything is about the destruction of what was previously a damned world, the meaning of "famine", "hunger" and "thirst" might not be that of destructive events per se, but rather spiritual famine, and hunger, and thirst. People who are left depleted of the belief in God, and who might feel the need for it, yet they can't comply.

I couldn't claim an ideal approach, but for the purpose of this book, I will farther refer solely to the literal meaning of the words – having to face global hunger and thirst. Could this lead to our doom? Could we actually all starve to death or die of dehydration?

Well, not really. I wouldn't dare call it impossible, but the probability of such a scenario taking place is very little. In and of itself at least. However, other events might, one way or another, bring them upon us as a consequence. We would be "alone in the forest" together. But, believe it or not, it would be much, much easier to survive in an actual forest than it would be in any of the following scenarios.

Precursors of famine

If you recall from our last chapter, we discussed how the dinosaur-killer asteroid smashed into the Earth and wreaked havoc on our favorite reptiles. But as we've pointed out, it wasn't the very impact that wiped out the dinosaurs. Not the fireball, not the shockwave and not the subsequent bombardment. While their effect can't be ignored, the resulted fallout was the primary stakeholder.

Debris was thrown into the atmosphere and beyond, giving birth to a giant layer of ash and dust that engulfed our planet, restricting sunlight access and leaving the Earth in darkness for years and decades. You know the story, so you can guess what happened next, hunger-wise. Almost all of plant life was gone. The water wasn't available to many either. It didn't freeze or boil, not that sort of thing, but large amounts of it ended up contaminated. Our planets inhabitants of the time were reduced to silence through thirst and starvation.

We could potentially meet the same fate as the dinosaurs, and it needn't be an asteroid. Any other cause with alike consequences could lead to the same outcome. So think "lack of sunlight" for a second. What else could pose the same threat? I don't blame you if nothing came to mind, but as we'll learn in the upcoming chapters, there are other blanket-type apocalypses besides that one.

A tremendous volcanic eruption is one of them. As you probably know, volcanic ash is a typical feat of such events. Volcanic material is thrown into the air, forming huge clouds of toxic ash which finally deposits on the surface. Most eruptions are not to be feared in that regard, but then some are truly apocalyptic. Without too much of a spoiler, the opaque barrier prevents light from reaching Earth's surface, partially or totally, which further lowers our chances of survival as clean water becomes a serious issue in the area and the agriculture suffers immensely.

Equally devastating (or even worse) would be the radiative fallout succeeding a nuclear explosion. Perhaps the leftover physical debris wouldn't be enough to cover up the entire planet, no matter how powerful the beast – multiple nuclear explosions might pull this off, though. Yet in this case there is another danger: radiation. In addition to food scarcity as crops die out one by one and all the other trouble caused by a nuclear device, water and air contamination would reach peaks unseen in the history of mankind. Thus, thirst and starvation may contribute to the horrifying aftermath of such a scenario too.

Last but not least, any other event that might culminate in or reach unbearable agricultural and water conditions would fit as a precursor of this type of apocalypse. Global warming accompanied by severe climatic changes is an appropriate candidate. So is another global ice age that we couldn't prevent with the current technology, whatever the reason behind it. There are many ways we could die of famine, no need to worry.

The exact cause and effect of these scenarios will be explored a bit later on. But in order to understand the issue and its severity, a more general assessment will be carried out in what's to come.

Running low

Seventy-one percent. That's the proportion of water on our planet. More than two thirds of Earth are covered in water. Even so, extreme water scarcity is an ongoing phenomenon which hinders over half of the world's population at least a month per year. And with the numbers getting increasingly unsettling and predictions looking quite terrifying, we ask ourselves whether we are running out of water.

On a planetary scale, the water is continuously recycled in nature. The water cycle keeps transitioning water from one form to another[1], influencing weather patterns and its overall presence in nature. Technically speaking, we cannot out of water, but disturbances in the cycle may result in extreme conditions.

With higher temperatures comes an increased melting of sea ice and glaciers, which puts back in motion water that has been previously locked away. Increased evaporation follows, leading to significant amount of water vapors being released into the atmosphere. Water vapor is a potent greenhouse gas, so it further increases the temperatures, which leads to more water vapor and so forth. This ill-starred causal nexus ultimately results in water scarcity. On top of that, higher amounts of water vapor might cause hurricanes which, unintuitively, empty aquifers as too much water arriving at once happens to simply run off.

But let's get back to our sheep. Water will be there for us, but do we need all that water? This is a whole other question with a negative response. As you may know, sea water and ocean water are salty – you can't really hydrate by drinking that. It actually makes things worse, so it can't be good for us. As it comes, almost all of Earth's water is like that. 97%, give or take. 97% of all water on Earth is undrinkable. Of the remaining 3%, less than a third is available in liquid form, with the rest being stored in ice caps and glaciers. So, it's these reservoirs depositing less than one percent of our planet's water that nearly 8 billion people rely on.

The little that we have is slowly replenished by the water cycle. The problem is that we are depleting it even faster. Whenever that happens, we recourse to our very finite reservoirs. Out of the 37 major ones on the planet, 21 of them are on track to be irreversibly emptied. At this unsustainable pace, we are getting ever closer to the firing line.

But how could it get to this? We do not consume that much water. Of course, you have a bath instead of a shower every now and then or maybe you don't close the tap while brushing your teeth, but other than that... one does not consume enormous quantities of water, not directly, anyways. There are a few other shrouded factors that take up much more that that.

Household usage is one thing, but then there are production expenses, industrial usage that is. Yet those average only 30% of humanity's fresh water consumption. Then comes the true colossus: agriculture. In some places, the situation is even more disproportionate, with agricultural usage going as high as 92% of the water consumption. All of that goes into crops and livestock, to ensure that the world's population is well fed. Remember how I said food and water are linked? – that's it.

Farmers are pumping water out of local wells until they essentially drain the existing aquifers. Apart from that, certain events toughen the process; whence the farmland calls for even more water supplies in order to make it through. This happens especially in arid areas, where water supplies are very shy in the first place. Those regions suffer from water shortages almost all year long.

With the growing population, chemical farming gained significant ground, as it is able to provide more and quicker. But with it came all sorts of pesticides and chemical preservatives. The effect those have on pollinators, especially bees, is unforgivable. By killing them or driving them away, our crops would no longer thrive. According to author Stephan Schwartz: "of the 90 foodstuffs that humanity lives on and requires to survive, 70 of them are pollinated by bees, and with the demise of the bees, the whole agricultural cycle is called into question and could collapse.".

As of now, 37-40% of the total land area is farmland. But, what if it was more? Despite looking dangerous to use that much water in the same scope, agriculture is among the most fundamental of necessities. We can't get rid of it, that's for sure. The workaround – we have to come up with better ways to administer it.

First, farmers should incline towards more water-efficient practices. Improved irrigation systems have the potential of crucially reducing water waste across the farmland. Over and above, it would save time, money and labor, while also preventing fungal diseases and weed growth.

Additionally, less gobbling crops could be put to use, so as to replace the fastidious ones. Based on the amount of carbon they consume per reaction, by and large plants split into C3 and C4 species[2]. C3s make up the wide majority of plants that we consume, while the C4s usually go to feeding animals (humans tend not to prefer them). And with agriculture slowly reaching peak productive capacity, the way to go would be to increase the plants' ability to convert sunlight into photosynthesis. With the existent C4s such as sugar cane, maize and millets, we observed that they are up to 300% more water-efficient than their counterparts. Through biotechnology, genetically modified organisms (GMOs) among crops could actually yield more resources by shifting from C3s to C4s.

Cover cropping is another reliable technique, through which specific crops are grown for the soil. It therefore helps for the building of better soil health, nutrient retention, erosion control and weed reduction. All that summarizes to less waste and more efficient usage of water and crops. Agroforestry[3], too, offers increased productivity, as well as greater diversity in the ecological goods and services provided.

Solving the issues with water and agriculture demands a multi-faceted approach. Water recycling techniques with proper filtration and purification need to be wide spread, whereas water pollution must, in turn, be well regulated. It isn't enough for factories and other companies to reduce their water usage, however. Those things must be implemented at an individual level. Reducing food consumption would directly reduce water usage. Choosing less water-intensive foods like red meat or nuts would contribute to the cause. Opting for a vegetarian lifestyle could save up as much as one third of your water consumption.

All of those small decisions could play an important role if applied on a larger scale. But with time, we will need more and more resources, so it's the techniques like the ones mentioned before that we must focus on. Most often, the water scarcity trend is fueled by poor distribution, management and misguided politics, more so than an overall global shortage. Same goes with food. That is in developing countries. In the developed countries, food and water waste is a more serious issue. There usually are sufficient resources, but over a third of them are wasted on average.

The resources are, or could be there, but it is our lack of inspiration that endangers humanity. Comedian Sam Kinison said: "You know what the problem is with world hunger? We've been sending them food.". Now, I don't know how much of it was meant to be a joke, but I find a deep and disturbing meaning to it.

Even so, however ignorant we might be and however greedy we might get with these precious resources, it still couldn't be called an apocalypse. Plenty will suffer, and the numbers will presumably increase even more with global warming and population growth, but it would not end us. Not today, not tomorrow. Yet the latter – population growth – might come with a slight surprise in the upcoming decades...

The Malthusian Crisis

The concept of "overpopulation" is nothing new under the sun. This idea refers to the point when a species' population (human population for all sakes and purposes) rises to such an extent which exceeds the carrying capacity of the ecological setting. In other words, it is when our population will reach a specific threshold, with the demand being overwhelming with respect to the production. A point when famine might be widespread, simply because we lack the productive capabilities to sustain all of us.

As I've emphasized, the problem with thirst and starvation isn't apocalyptic in the present day. It is a serious global issue, but it's theoretically solvable. We have the necessary resources and technology to sustain the human race. In an overpopulated world, this wouldn't hold.

If we were to eventually get to a state of overpopulation, things would only go downhill from there. We might have all it takes now, but as human population increases faster than we can adapt and advance our production, many will live to an undesirable destiny. It is generally considered a one-way ticket: you can easily get there, but there is no return unless some sort of catastrophe which drastically reduces our number takes place. This is precisely what Thomas Malthus predicted over 200 years ago.

He believed that the human population will skyrocket in the future and that it will reach a point like this, when it will far surpass our planet's agricultural production capacity. Once that happens, we might face sharp reductions in population as people are unable to provide subsistence produce for themselves – many will die, that's what he meant. And at the forecasted population levels of that time, the death toll could be so severe that it may well be considered apocalyptic. This was later called "The Malthusian Catastrophe" or "The Malthusian Crisis".

It was said that the population would increase exponentially with time for the first part. Then, sometime after the limit is passed, severe drops would occur, perhaps just as abrupt as the initial increase. It might then stabilize again, but still tend towards an increase. After that, another drop would happen and so on. The population vs. time graph would look something like the teeth of a saw, with spikes and troughs, some more severe than other.

Accounting for the agricultural advances and the new methods of producing and providing food that might come up, the graph could be a bit inclined. The teeth would get smaller and smaller, until some sort of stability point is reached. The oscillation would still be pretty harsh, but the population could stabilize in the end, at least in principle. Even so, the initial drop, which we haven't reached yet by the way, would be particularly extreme (the others might not be such a piece of cake either).

Back in 1798, Malthus stated in his work "An essay on the principle of population": "The power of population is so superior to the power of the earth to produce subsistence for man, that premature death must in some shape or other visit the human race. The vices of mankind are active and able ministers of depopulation. They are the precursors in the great army of destruction, and often finish the dreadful work themselves. But should they fail in this war of extermination, sickly seasons, epidemics, pestilence, and plague advance in terrific array, and sweep of their thousands and tens of thousands. Should success be still incomplete, gigantic inevitable famine stalks in the rear, and with one mighty blow levels the population with the food of the world."

Quite macabre, but he might not be mistaken... Remember that that was back at the end of the XVIII century, when the world's population was below one billion. A century later, it reached 1.5 billion. Yet just one more century further do you know how high it got? Just a bit shy of 8 billion people! 8 billion! Just to see how high is that, try counting to it. I'm kidding, don't count to 8 billion. If you were to count one number every second, nonstop, it would take you more than 3 lifetimes to reach that number.

But it's not only the number itself that is astonishing. It is the difference between that of just a century ago – less than a fifth than what it is today. Back in the day, Malthus might have imagined that 8 billion far exceeds the overpopulation threshold, but we're still handling it. For how long? that's the question.

In the last century, the massive spore in population numbers across all continents was due to the medical and technological revolutions. Most notably, the discovery of antibiotics, vaccines and other advances which substantially reduced the main causes of death from the past. The increase was pretty natural. Just as helpful was the improved agricultural techniques and the novel technologies in the field. This was also given a name – "The Malthusian Trap" (improved agriculture further increases population which makes the demand even bigger).

And it looks like the population trend keeps going. But we are not likely to stop developing ourselves just to avoid overpopulation.

Contrary to the expectations, scientists believe that we will stabilize ourselves sooner rather than later, in spite of the ever more favorable living conditions, even before reaching the threshold. However, no certainty exists. Most predictions describe a population of 9.5 billion by 2050, or 11 billion by 2100, but nothing is clear yet.

Unpredictable changes in population and demand happened in the past, and there is no guarantee that something similar will not happen hereafter. For example, decades ago, China was a major exporter, far outstripping imports. But a huge spike in population meant that the country was no longer self-sufficient, relying on imports more than any other in the world. Africa presents another issue, with its demand growing nearly as fast as China's. While China and India are currently the most populous countries, their population shows some early signs of stabilization. In contrast, the continent of Africa is experiencing an enormous rise in population.

Before very long, China, India and the African continent combined might make up more than half of the world's population. It isn't the eye-catching numbers, though. It is more about whether these regions are self-sufficient or not. Out of the three, India is doing best in this regard, with its demand not topping internal production. This was largely due to the Green Revolution[4] that took place there, something which no African country has yet experienced. Maybe this is what's needed for self-sufficiency.

Interestingly, experts argue that not even this will help certain regions. Notwithstanding the Green Revolution in India, it is still predicted that the population and economic growth will further increase demand, rendering it a net importer of calories. Same applies to China, which is already in a pretty rough spot, and soon Africa as well.

Until now, even the countries which heavily relied on imports were thriving because of the productive capabilities of the other regions. But this will become increasingly challenging as more and more countries take the same route. Europe and the Americas will provide the main sources of calories, with South America being a huge step ahead of the others. This will come at the cost of massive deforestation.

In this dystopian world, we are left with hundreds of trillions of calories deficit, translating to millions and billions of people suffering from extreme scarcity, doomed because of lack of produce. And that all while assuming that all the exporters are providing solely for the 3 major importers mentioned before – China, India and Africa.

Conclusion

Avoiding this kind of semi-apocalypse isn't easy, but it's doable. Note that I call it a semi-apocalypse due to the rapid outcomes that might ensue, which wouldn't let us all go extinct. It is some sort of failsafe mechanism: saving humanity at the expense of a decent chunk of it.

The resources of our planet are limited too, so we could also count this worry in. But until we are capable of fully exploiting them, there is still a long (though not very) way to go. We shouldn't be concerned about that for now – we will find methods of dealing with that in the future.

Regarding the unignorable overpopulation scenario, it should be much more frightening than that. If it were to happen in your lifetime, perhaps you wouldn't suffer as much, but think of all those people that would. This doomsday event was ranked so high simply because of the intimidating trends that point towards it. Yet that also gives us some spare time to prepare and even avoid it completely.

Solutions like changing consumption patterns and reducing food waste are viable options, but they haven't been given much attention lately. Those would imply the prosperous, exporting countries to change their behavior in favor of the deficit regions. And like Sara Menker (founder and CEO of Gro Intelligence) highlighted in one of her speeches: "Waiting for others to change their behavior on your behalf, for your survival, is a terrible idea."

The optimal way to proceed would be to reform and commercialize the industries of the not self-sufficient regions so as to correspond to the growing needs of the people. Improving agricultural practices and infrastructures worldwide is also imperative for the well-being of our species. Adding to the equation the individual efforts to combat these issues... we realize that there are, in fact, a number of solutions to prevent this gruesome scenario. We just need to take action!

About 40% of the Earth's total land area is covered in farmland, but the percentage may soon grow as population rapidly increases

Genetically modified plants are an innovative solution to the high demands in agriculture

1. Water changes state, from solid in glaciers and ice caps, to liquid in seas and oceans, to vapor in the atmosphere, then rain and so on.
2. C3 and C4 plants function via different mechanisms. The former have just one CO2 acceptor and perform photosynthesis only when the stomata are open, whereas the latter have 2 acceptors, and perform photosynthesis even when the stomata are closed, rendering them more efficient.
3. Agroforestry is the intentional integration of trees and shrubs into crop and animal farming systems to create environmental, economic, and social benefits.
4. A Green Revolution refers to a large increase in crop production in developing countries achieved by the use of artificial fertilizers, pesticides, and high-yield crop varieties.

III.4: Climatic changes

"Our house is on fire..."

Greta Thunberg

Imagine you are quietly sitting at your place, lying in bed and watching TV with your friends. Then, all of a sudden, you smell something funny. It's as if something is burning, but you don't bother get up and investigate. You begin to see smoke coming from the doorway, but the movie is so good, you might just wait a bit until it's over.

Damn commercials! Now there is no reason to stay in bed, so you think you should check the house and identify the source of the smoke. But your friends wouldn't let you get up. You fall prey to their begging you to stay, only to see fire engulfing the room. You start panicking and so do your friends, yet the movie is so close to being over... Before you know it, the fire burns down the house, with you inside, and you never managed to see the movie finale.

"What a crazy scenario! That would never happen in real life!" – you say. If only you knew that the real situation is much worse... Of course one would go check the smoke the moment it makes its presence felt. You might just be facing a small malfunction of an outlet or something similar. If you have a fire extinguisher, putting out the fire is child's play. Even if you somehow miss the early signs, say you are sleeping, whenever the fire becomes obvious to you or the ones around you, the firefighters would get there in the blink of an eye.

Now think of this big house that is our planet. You and your friends are the entire world's population. You smell the smoke – realize that so many of our actions are wrong, that we are polluting non-stop and that we are so reckless in our everyday lives. You see the smoke coming – all of the negative effects that show up as a result of our harmful activities. You even see the flames spreading to you room! – temperatures rising, the nature going haywire and people dying. You still don't do anything to stop it.

How does this sound? The entirety of Earth's human inhabitants in a single burning house and doing nothing against the flames. We are so captivated by our ever-emerging issues and conflicts that make up for such a good movie, that not even the fire will stop us from watching. Commercials may pop up every once in a while, with scientists and activists warning us about the imminent threat, but the society pulls us back, preventing any further action which has the potential of saving us.

With the fire at our door, we have a last chance of putting out the fire ourselves, because unlike in the aforementioned hypothetical scenario, there are no firefighters to rescue us. It's all on us. We started the fire, and we must end it before it burns us alive.

"The greatest threat to our planet is the belief that

someone else will save it"

Robert Swan

Disambiguation

Remember our good-old friend, the atmosphere? As it happens, it consists of many elements besides the usual oxygen and nitrogen. And we've highlighted the importance of them all in past chapters – whether it be the breathable air, the burning of meteors or the shelter it provides from radiation, the atmosphere is paramount to our survival. But there is still one role we didn't cover.

One of the most evident purposes of our atmosphere is keeping Earth's conditions appropriate for the development of life. The previously mentioned Goldilocks zone of our Solar System technically doesn't even include Earth. Even with the albedo[1] of 0.36 (which is, by the way, also due to atmospheric composition), we are still too far away from the Sun to be considered a Goldilocks planet. That means, the temperature which would arise from calculations wouldn't be high enough to allow for liquid water, it would be twenty-something degrees below 0 Celsius.

But then the Sun doesn't emit equally at all wavelengths. Similarly, the Earth's atmosphere doesn't absorb equally all throughout the spectrum. For instance, it absorbs much more visible light than infrared. This is due to the particular composition of the atmosphere. It also doesn't let as much surface radiation to escape back into space. Taking this into account, we realize that above-zero temperatures are, indeed, possible on our planet (as we can see). At the same time, we also understand which gases have what effect on the retention of solar radiation.

The phenomenon thanks to which we thrive on our warm planet is known as "greenhouse effect". Subsequently, the atmospheric components which maintain this effect are called "greenhouse gases". The most important of those are water vapor, carbon dioxide, methane, nitrous oxide and chlorofluorocarbons. Those have been here for hundreds of millions of years, circulating in the world, much like water does. But something odd is going on with them in latest years.

Like Elon Musk presented the issue, "the thing that's changed is that we've added something in the mix – this is what I would call the turd in the punchbowl". He was talking about the extra carbon[2] that has entered the cycle; carbon that had been buried for such a long time, but which got unleashed with the help of human activity.

Looking back, there have been periods with excruciating carbon levels fueled by major natural events like volcanic eruptions. But at no point were there levels so high as they are today driven by ordinary activities. "Ordinary" isn't really the case, though. Today's world is strikingly different than it used to be millions of years ago, before humans appeared. It is still very distinct even from earlier human times. Spectacular changes can be observed if we peek at the world only a few centuries ago.

So, what changed? The technology did. It took us hundred of thousands of years to evolve to our modern selves, and only a hundred to make an exponential jump to the current technology. This last century was, no doubt, the most remarkable in all of history. But, as I've said, this wasn't great from the viewpoint of carbon emissions.

With the Industrial Revolution and the major technological boost, we learned to properly exploit the environment's resources. This plethora of anthropic actions led to terrible levels of carbon in the atmosphere, many times higher than in previous centuries, even decades! And, intuitively, higher greenhouse levels result in higher light retention and thus increased temperatures across the globe. This is how the term "global warming" was coined. The increased temperatures also trigger other sorts of natural problems, falling under the name of "climate change".

This is how we could burn alive. The difference in temperature between today and 1880 is estimated at 1.4 degrees, with 1.1 of those coming from 1975 on. Worse still, the next 10 years might bring about another 1.5 degrees increase in global temperatures. Carbon levels are, scientists say, at the maximum in at least 800,000 years.

The one, two or three degrees in change could be seen as insignificant, but the implications are enormous. Not only could we burn alive. We could be burning... while starving... while also drowning. The scenario gets grimmer and grimmer the more we understand things. But we don't seem to. Doctor John Rilley of the MIT Joint Program indicated that our problem is that "we don't understand how life is all interwoven and what pulling those threads out might do.". One scenario could easily spark from another and, in turn, give birth to yet another. This is how apocalypses are created.

Embracing the flames

A famous saying of the Ancient Greek philosopher Heraclitus was that "the only constant is change". Be it good or bad, change is assured. In the grand scheme of things, change has hitherto acted to our advantage. But what if it's overdriven? Something as unpredictable as us could easily turn the course of things upside down. We are especially skillful when it comes to climate change.

We are altering our surroundings in unforeseen ways. Ways that do not coincide with the planet's needs. And while Earth is a great storage of valuable resources, it is also a home to us and to all other species that live alongside humans. Through our egocentric disposition, we are exhausting nature itself, leaving no room for harmonious living, and giving way to abnormal effects.

Causes are many – pollution (air, water or soil pollution), deforestation, the burning of fossil fuels and the general displacement of nature's elements, you name it. They all play their card in the changing patterns of Earth. One way or another, they interfere either indirectly, via the carbon cycle, or directly, with the world's wildlife and ecosystems. You can break it down to "hasty development". That's what we do. And that's how we influence our entire planet with so much ease.

I assume you've heard this over and over again: that pollution is bad, that more and more forests are being cut down or that cars and factories and what not emit gargantuan quantities of carbon dioxide and other gases. For this reason, I will not bother detailing every bit of that to scare the life out of you. I will, however, emphasize some of the already visible consequences of this unreasonable way humans are carrying things out.

A short story my dad told me once summarized it best for me. Once again, it involves my home town, earlier described as hosting one of the biggest ecological disasters ever. As described, there was also a period in back in the 80s when pollution reached an all time high. Not because of cars or waste – people were pretty conservationist in the communist era. It was all because of the enormous metal plant situated right in our city. He remembered how, as a child, he couldn't play much outside during certain periods and that he could taste the lead in the air.

Just imagine that: children (and everyone else for that matter) not being able to spend time outside because of exceedingly high pollution. It really touched the younger version of me, as I was about the same age when my father told me the story. But this is the reality we live in. Thankfully, our city is no longer host to the highest pollution levels in the country, though it's not very far behind. But sadly, hundreds if not thousands of cities around the world suffer from the same reason, particularly highly populated ones in China, India, Brazil and even the US. The more demand there is in the world, the more "hasty development" we face, and so pollution clouts, forests are cut down and large arrays of ecosystems become devoid of life.

Climate change, overpopulation and the growing issue with food and water shortages are tightly linked. It's as if we're stuck on a crash course with little chances of survival. Barriers obstructing the way and obstacles hitting us from all sides. Because it's not just the abominable temperature rise. It's also the more frequent droughts, the stronger hurricanes and the monster floods. An estimated of more than 150 species go extinct, on average, every day! Not only do we lose the biodiversity and natural habitats, but we are also further toughening the already worrisome situation with food production.

Former US vice-president and Nobel Peace Prize winner Al Gore explained how "foods and crops are optimized for the climate conditions that we are now changing". If we continue to modify these conditions, eventually, our crops will no longer be able to keep up and prove inefficient in a time of most need. On top of that, more water will be required for the free-falling industry of agriculture. We are currently in the process of developing more sustainable, less pretentious plants to yield better produce, but if we mess with the existing crops' natural compass, we have no hope to work something out in time.

With changes in temperature and natural phenomena which will definitely pose a problem for years to come, changes in the water circulations cycle will commence as well. This will lead to all sorts of threats to our society, but probably most noticeable will be the overall increase in sea levels. As the Earth's ice caps are swiftly melting, we could expect changes of a few meters in sea levels.

At a global standard, this is definitely not much, but from a societal point of view, it is catastrophic. A number of metropolises and conurbations are located near open water bodies such as seas and oceans. With just a slight couple-of-meters increase in sea level, entire countries would become submerged – another Great Flood scenario, only permanent and not as abrupt. Billions of people would need to relocate and the population of some megalopolises could soar into the hundreds of millions. We should be able to overcome this, we've dealt with migrations before (not as big, but still). We are also warned well in advance so there is time to improve and, worst-case scenario, flee the place. But so far, the danger only came from indirect sources. It's about time we saw how global warming could directly impact us.

Predictions would indicate a global 3 to 4 degrees increase by the end of the century. We are rapidly warming our planet as we are speaking, but it is believed that the trend will normalize in the following decades, and that our impact wouldn't be as steep. Either way, even if it will be just the end of the century which culminates in such temperatures, the effects would be devastating. Senior researcher at the French National Center for Scientific Research Gaël Giroud, for one, doesn't see things to brightly. He claimed that "maybe some images we find in the last book of the Bible, the Apocalypse, will have to be interpreted literally 74% of people by the end of this century to be under lethal threat because of heat; so I would not expect the entire mankind to disappear, but certainly genocides to take place".

Images like "A third of the earth was burned up, a third of the trees were burned up, and all the green grass was burned up." (Revelation

8:7) do somewhat resemble the story of global warming. But the reason why most of us wouldn't make it too far in this "hot world" is more complex.

We wouldn't burn alive to begin with. However, certain regions will be severely impacted by the rising temperatures. The flora in those areas would perish, thus rendering them uninhabitable. Humans and animals alike will migrate towards more suitable regions. The only problem is, there are very few of those left.

With a mere 4-degree increase in temperature across the planet, about half of all the land area would become deserted. Everything less than 45 degrees from the equator (latitude degrees this time around) would be pure wasteland. The only suitable regions would be the ones closer to the poles. Africa would be completely inanimate (no habitable areas available). Of Europe, Asia and North America, only the northmost regions would be appropriate. These refer to the Baltic states and the UK, most of Russia, especially Siberia, and Canada, alongside with the northern states of the US. Of the Australian continent, the populous regions of today wouldn't be too permissive when it comes to human habitats, but Tanzania, the Southeastern Australia and above all New Zealand would make up the new havens for humans. South America would meet a similar fate to Africa, with only the extremity of Chile being life-sustainable. It may come as a surprise, but significant portions of Antarctica and Greenland would also suit use in that time of crisis.

Not too shabby at first sight. But think of an overpopulated world, with enormous demand, forced to reestablish and essentially redefine itself, as the formerly prolific territories become irrelevant and unsuitable for our lives. The change would be radical and large numbers of people wouldn't succeed in such a place.

If all that wasn't enough, there is also the notorious relationship between humans and microbes, as Al Gore also mentioned. That is also quite sensitive to temperature. Changes of these proportions might very well spark new variants of viruses which are currently ineffective or even allow for new, deadlier ones to reign over our world.

An apocalypse like this is way more probable than an asteroid killing us. That is simply because we can already see it coming. You know asteroids aren't going to hit or that they might pose some threat, but the threat of this is so great that it can't be ignored. That's how it should be, anyway.

We ought to prepare and change our activity also with respect to the environment's needs as soon as possible. Otherwise, we could really get to that point of imminence. Researchers the world over keep talking about a "tipping point". Again, a point of no return, after which there is no hope for our civilization. But how true is this? Should we be skeptic?

Fact or hoax?

The main argument for the existence of such a point is the spectacular level of atmospheric gases we've managed to reach in just a few decades. Sea level rise, variations in weather patterns, and even the more extreme events like droughts, hurricanes and floods are all "live" evidence of how climate change is hurting our planet. You've certainly heard more and more news talking about how often these events became. But are we that close to the tipping point?

UN Secretary General Antonio Guterres thinks so. He called it a "code red for humanity". He went on to say "The alarm bells are deafening, and the evidence is irrefutable: greenhouse gas emissions from fossil fuel burning and deforestation are choking our planet and putting billions of people at immediate risk.". The Intergovernmental Panel on Climate Change (IPCC) also gave disturbing reports regarding our persistence to stay on the wrong path. While others disagree with these reports, claiming that not only are the estimates too harsh, but also that a tipping point doesn't exist before all else, our eyes cannot be tricked as easily.

It's one thing to trick our minds through controversies and conspiracies, but when we begin seeing this mountain of evidence with our own eyes, we can safely assume that something is going on. Yet with all those news and scientists and environmentalists stressing the issue, a concerning amount of people straight-up deny climate change.

As a parenthesis here, there was this analogy with aliens that not believing in their existence is like taking a glass of water from the ocean and claiming that there are no fish out there. But as far as aliens go, we do not have concrete proof of their existence, whereas we know for a fact that fish do exist. I think this analogy would work better in the case of climate change. Like with fish, we know climate change is real. It is proven and there is concrete, palpable evidence, if you may, that the world is changing this way because of us. But then one looks up and sees no smog, notices it isn't difficult to breath, the thermometers don't go wild from day to day and there are still trees around. At that point, one claims there's nothing wrong with Earth, nothing wrong with us – climate change isn't real. All the while people are suffering from pollution and faulty natural phenomena, over 40 million trees are being cut daily, and seaside settlements are losing ground in front of the merciless ocean levels.

"What about animals? What about us? What about the seas? What about killing fields? What about cutting trees?" So many questions unanswered and left into oblivion by our ignorance. Honestly, if not even Michael Jackson with his "Earth Song" can't convince you that climate change is real, I don't know what can.

Actually, there is one thing that beats them all: misinformation. Usually it takes the form of manipulation, also known as purposely distorting the evidence such as no one notices how wrong they are, and the ones that know just count it as a mistake – that's about right. We see that all the time, to the greatest degree in the world of politics.

So many powerful people disregard climate change as a major issue from whatever reason. Maybe they could benefit from that, or maybe they just refuse to recognize it and focus on their own problems. This type of selfish behavior creates confusion among the masses and they practically become buried in the sand. They see the drastic changes, but trust other, "more knowledgeable" persons above themselves.

I, for instance, hear a lot of remarks on seasonal changes from my older relatives. Thigs like "wow, when we were young there was snow every winter, not like today when it barely snows once a month" or "in the time of so and so, the summer only came in July, not as early as March like nowadays...". And it can't be just me for sure.

But then come oil companies or glorified politicians that turn a deaf ear to all that means climate change and global warming. Are you kidding? Countries withdrawing from environmental accords as per the fact that "the measures imposed are unreasonable", or that "it wouldn't make much of a difference even if we all stayed in, so we just decided to drop out". Companies promoting their products as being eco-friendlier to increase their sales, while continuing to produce and therefore maintaining the damage dealt. Why do those environmental conferences even exist if none of the important lawmakers and action takers implement the measures advocated by scientists and environmentalists?

In a pickle

It is crucial that we understand the serious jeopardy we are confronting with. The tipping point might not be as close, but we may not be that far away from it either – it does look like the situation could become irreversible. We are also not sure how bad the outcome would be, but we'd have to consider the worst scenario possible to prevent it from happening.

We can act collectively and strive for a brighter future. Cutting waste, reusing and recycling as much as we can, and, most importantly, opting for sustainable products. But the issue is greater than that. We are in very big of a pickle as the ones in charge refuse to take measures as to avoid this scenario. Companies repudiating the idea that climate change will bring doom upon the environment (which we are part of) and policymakers declining any suggestion that comes from the scientific or activist community.

What is more, they are thriving in the world that they are wounding most. In fact, the richest 10% globally account for half of all emissions. As climate journalist Jelmer Mommers presented it "the western countries got rich, while causing climate change, and now rich consumers are causing more climate change". For reference, the US and Europe make up 52% of the world's emissions. And who does this imbalance affect most? – the poorer regions. This could be another motive why climate change is widely being ignored: it doesn't affect the ones with decisive power. But don't worry, it soon will.

Starting with the poor countries and increasing poverty all over the world, climate change will act in the cruelest way imaginable. Though it is predictable. We have to make the most of that. We have to focus our strength into this last ditch to halt, or at least dampen climatic oscillations to the point where they will stabilize. This is our go-to for the following decades. But we have to start now!

Share is care in this context as well. Promoting environmentalist ideas and choosing more sustainable alternatives anywhere from household appliances to our everyday lifestyle would really help fight against this gruesome enemy that is killing our planet. But just that might not be enough. We cannot count on everyone to take initiative, so we'll have to urge our lawmakers and world leaders to take more nature-appropriate approaches all throughout society. As soon as possible that is, otherwise your children might have to grow gills or sweat sunscreen (depending on where they live).

Conclusion

There is no time to give up! Many have this flawed mindset that if changes aren't straight-forward they're not worth it. But nothing is straight-forward, and these changes are the most worthwhile. It does take collective effort to drive the change, but if you decide to not join the fight, then how could you expect others to do it for you.

Very motivational, I know. Not everything in this book is meant to be sorrowful. Still, this kind of apocalypse is real. Think what you want, but if it hadn't been, it wouldn't have made it this far up. The entire Book of Revelation focuses on how humans have destroyed their world and themselves as a people, and about how a drastic change is needed for atonement. It is exactly what is happening to us, but we do have hope. Hope that we wouldn't get to the doomsday point and that we would make amends to our injured planet just in time to turn things around.

The strong connection between this type of scenario and some of the aforementioned ones may further accentuate our hurry. All in all, we are facing the biggest challenge of this century and we don't have any spare time left to be kicking our heels. Things are already bad enough and only you, and me, and the rest of us can be the change...

42 million trees are cut down daily; the Amazon rainforest is already 17% deforested, with estimates reaching 27% by 2030; scientists believe that the tipping point will be reached at 20% to 25% of deforestation even if climate change is tamed

Pollution levels have reached an all-time high in many regions of the word, especially the ones also dealing with the problem of overpopulation

Glaciers are melting at an ever-faster rate – this could lead to a one-meter increase by 2050, and 3 meters until 2100. Large portions or even entire cities and could be underwater just a few decades from now. These include Tokyo, Mumbai, New York, Osaka, Istanbul, Kolkata, Bangkok, London, Ho Chin Minh, Amsterdam and many others

1. The albedo of a body is defined as the amount of incident (incoming) light which is being reflected by its surface. An albedo of 0.36 out of 1 means that 36% of incident light is reflected back into space by Earth, while the rest 64% (in our case) passes through the atmosphere and reaches the surface.
2. When talking about carbon or "carbon emissions", we refer to all carbon-based elements which have the important retention power that keeps our planet warm.

III.5: Solar activity

"But the day of the Lord will come like a thief. The heavens will disappear with a roar; the elements will be destroyed by fire, and the earth and everything done in it will be laid bare. [...] as you look forward to the day of God and speed its coming. That day will bring about the destruction of the heavens by fire, and the elements will melt in the heat."

2 Peter 3:10-12

Praise the Sun! For we wouldn't be here without it. This diurnal ruler of ours does everything in its power keep shining, to keep nourishing us with its warm rays and cosmic protection. It is thanks to the Sun that our planet is so fine-tuned for life and everything that exists down here. That bright fleck keeps Earth alive.

I've repeatedly stated how important of a role our Sun plays, but something like this cannot be overstated. It is the symbol of power, of grandeur, and also of kindness and virtue. In many ancient religions (but also modern ones), it was considered nothing less than a god. People understood its significance and deemed it valuable above all else. In the Chinese culture, it represented authority and dominion. The Zia Indians of New Mexico view it as a representation for the circle of life – usually portrayed as emitting in four directions, this bounds to the changes undergone throughout life. The Egyptians had the well-known Winged Eye symbolizing divine knowledge and the Aztecs had their own "Eye of God", thought of depicting the coronas seen during solar eclipses.

In most cases, it is related to hope, to renewal and the cyclicity of life, driven by its eternal dawn to dusk motion and beyond. But just as the Sun can be protective, it can also exert its power in more mischievous ways. After all, there's more to it than it gives the impression. The Sun is a literal ball of fire and plasma; gigantic in size and not that far away from us. It is the definition of chaos in our Solar System. How would you imagine a colossal fireball behave all the time?

So, as we've only hailed the Sun in our apocalyptic discussions, we shall also explore how the Sun itself is very much capable of producing a doomsday-type event at any given time.

Into the maw

You see, the Sun isn't just a big ball of gas. I mean, it is, but there are a few internal processes going on there to whom this star's greatness is due.

While it seems to be pretty solid from afar, the Sun is more of a tremendously hot ocean. So hot that it rips electrons and nuclei apart, leaving them freely flowing around each other in a state called "plasma". This plasma consists of ionized gas – that is, positive ions and free electrons[1].

Because of the very high temperatures, all of these particles gain substantial kinetic energy. Combined with their intrinsic electromagnetic energy, at close range (of atomic sizes), they merge together, giving birth to other, more massive elements, releasing energy in the process. This phenomenon is known as fusion, and takes place in the inner layers of the Sun, accounting for its energy output. The results of this process also take the form of plasma. So, per general, all of the Sun's little components aren't strong enough to withstand the internal temperature, and give way to the formation of plasma.

Plasma is pushed around and shaped by the Sun's magnetic field. Of this magnetic field: we know how strong it is, capable of protecting the entire Solar System from interstellar radiation, and also that its formation mechanism is similar to Earth's magnetic field, only on larger scales. But unlike other fields in the universe, the magnetic field comes with a twist. It is part and parcel of the electromagnetic force. But so is electricity. Therefore, a magnetic field will always be accompanied by its counterpart, an electric field, and vice-versa.

The electrically charged protons and electrons create a magnetic field as they move, and this magnetic field then shapes the pathways of particles. Like inside Earth's core, we deal with a dynamo effect, which keeps the particles moving, while producing a magnetic field, which further drives them, thus maintaining the Sun's powerful magnetic field. As it stretches so far out, this field stores enormous amounts of energy, in the form of plasma. This leads to constant trickle of solar plasma to be carried around the Solar System, as it leaks out. The farther you are from the Sun, the lower the magnetic field and the milder this sort of rain, called "solar wind".

But as it happens, winds here on Earth aren't very organized, are they? They come sporadically and with various intensities. Space winds are no different. There isn't a steady flow of solar wind going in all directions, everything is chaotic. Really, the Sun is more of a jumbled junction of all sorts of things rather than just a shining fireball. Over and above, the magnetic field – kinked and curled and what not – messes with the traffic lights of the already crowded junction in unexpected ways.

The insides of this celestial beast only manifest themselves truly as they come out of the maw...

Of bursts and flares

The fact that the Sun is around 150 million kilometers away doesn't stop it from shooting all kinds of radiation, millions of tons of superheated plasma and charged subatomic particles our direction at significant fractions of the speed of light. All of these extreme space weather examples are caused by alterations in solar behavior. Thinking about it, it's unsure whether the Sun wants us dead or alive.

While magnetic fields are tangled, stretched and twisted the year round, for the most part, solar activity can be considered mild. However, there are certain periods when the Sun wakes up spewing other-worldly material towards us, before going back to sleep. We don't know when it will burst, but we know when to expect it. "The Sun is like a volcano", said a former director of the NOAA's Space Weather Prediction Center, "we can't predict when it's going to erupt but we can see the signs building up".

From a physical standpoint, we can infer that the maximum build-up takes place at the time of pole reversal. Unlike Earth's magnetic field, the Sun's field is way more volatile. Yet the story is unchanged: it undergoes the same type of polar reversal event every now and then, the difference being that this "now and then" is more frequent than in the case of our planet. It's not several hundreds of thousands of years, it's just eleven.

About every 11 years, the Sun's magnetic field does a flip. In other words, the north pole becomes the south pole and the other way around, but you know how it goes. This flip is one aspect of the roughly 11-year activity cycle the Sun experiences as its magnetic field evolves slowly over time. As the cycle progresses, the Sun's stormy behavior builds to a maximum, and that's when the magnetic field reverses. Then the Sun settles back down to a minimum, only to start another cycle. Hence, it is around the time of magnetic field reversal when we should expect the most intense activity.

The whole cycle takes double the time – that is 22 years before the poles get to their (approximate) original position. The half-cycle corresponds to the periods of maximum intensity, also related to the biggest sunspot presence at the surface of the Sun.

Sunspots are areas of particularly strong magnetic forces on the surface. They generally appear darker because of the lower temperatures associated. These lower temperatures are generated by the same magnetic fields, which are so strong that they keep some of the heat within the Sun from reaching the surface.

A peculiar feature of theirs is that they always come in pairs. Each of the two components of a given pair constitutes one end of a magnetic knot. It's as if a magnetic loop originates from these two areas. Inside those knots, enormous amounts of energy build up. Sometimes, they just deflate like a balloon as the period of maximum intensity (and maximum sunspots) passes by. But other times, the energy is so great, that the knots break apart. The Sun consequently vomits plasma and other energetic particles into the Solar System, clearing everything in the path.

One type of projectile the Sun occasionally sends at us are solar flares. These are, as the name implies, brief bursts of energy. As the magnetic field goes crazy during times of maximum activity, magnetic fields sometimes break and point outwards. The charged, energetic particles still obey their trajectory and they blown outwards by the distorted field lines. The constituent particles are accelerated at high speeds, carrying powerful radiation into space.

Another example of violent solar activity – Coronal Mass Ejections, or CMEs. These huge bubbles of radiation usually form through a sunspot rupture, as mentioned above. This time, the magnetic field breaks as a result of the high energy stored, but it then reconnects and chills out. CMEs release billions of tons of material racing through space at more than a million kilometers per hour. In contrast with flares, the speed of CMEs is quite small compared to their magnificent sizes (they get bigger and less dense as they recede from the solar surface, dwarfing the planets they face in their way). Also, they can reach far greater distances than solar flares can.

This dangerous duo of solar storms is considered the most violent of explosions in our Solar System, releasing the energy of billions of hydrogen bombs. But in all their infamy, solar flares and CMEs usually have no impact on our lives. Even when they are stronger in periods of high activity, the best they do is slightly intensify our auroras as they are directed by the Earth's own magnetic field.

Having a more exquisite light spectacle in the polar skies can't be bad for us. Who wouldn't wish that? But you should be careful about what you wish. If you want to see the beautiful auroras, I recommend visiting the polar regions and not waiting until you see them elsewhere.

Carrington events

There were some people throughout history who didn't visit any of the poles, yet they did manage to see aurorae in lower, even equatorial regions. For most, it was just the basic: eerie, bright traces in the skies, of unearthly origin, that surprised them with an unexpected light show. Others were not so lucky...

This happened back in 1859, and it was the result of the most intense solar phenomenon recorded. Later dubbed "the Carrington event", this solar storm of unforeseen proportions happened during the midst of a solar cycle (obviously) – solar cycle 10, to be more specific. And because of the sheer strength of this storm, something known as a "geomagnetic storm" is most commonly used to refer to this type of event.

Once or twice a century, storms of much higher intensity are born from the Sun's chaotic magnetic fields: solar super storms. The earliest signs of an upcoming super storm are spontaneous flares in the solar atmosphere. During the increased activity period, these flashes would occur first, foreshadowing the more treacherous thunder. Of course, it wouldn't be an actual thunder. It would be a massive CME, travelling at menacing speeds. A CME like this would make its way to Earth's orbit in about a day.

If it were to actually meet Earth in its path, things would quickly aggravate. When it is to arrive, it would cause a violent compression in the Earth's magnetic fields that has been all protective until then. Fully compressing it to the point that the storm gets to the surface is highly unlikely, but we will discuss that a bit later on. Almost never does this happen, but it could be that the aligned CME's field coalesces with ours, elongating Earth's field into a tail as it passes over. Like in its formation, the energy of the CME, now contained in the joint field, becomes to large to fully embrace. This resulted field snaps as well, causing the energy stored to shoot towards Earth. The alarming process is what scientists call a "geomagnetic storm".

Now, something wouldn't burn us alive – we still have the magnetic field to look after us, however distorted. But you know what would be burnt to crisp? That's right, all of the circuits of the electrical components would stand no chance. It is the same dynamo effect in action. The charged particles in the storm would interfere with their kin here on Earth, twisting circuits in dubious ways, and either shutting them down completely, or even destroying them.

When there's nothing to toast, a geomagnetic storm would barely be noticeable if it weren't for the strange auroras. This is exactly what people experienced some century and a half ago. There weren't many electronic devices in 1859, but none other than the telegraph was the talk of the town. At the time of the Carrington event, all working telegraphs were, accordingly, either shut down or destroyed. Some of the circuits burst into flames or exploded, shocking and injuring those nearby. Quite a few deaths took place during the geomagnetic storm, but in general people were doing fine and quickly recovered from it.

The month of March 1989 culminated in a similar event. It was smaller in scale, but had a bigger impact on society as technology was more advanced. The entire province of Quebec, Canada, suffered an electrical power blackout. After darkness engulfed Quebec, bright auroras spread as far south as Florida, Texas, and Cuba. Reportedly, some onlookers thought they were witnessing a nuclear exchange, while others thought it had something to do with a space shuttle that was launched the same day. Actually, it was just the misbehaving Sun deciding to play a prank on Earth. And it wasn't even at the peak of solar activity during the respective cycle, just a CME that came out of nowhere. The day of March 13th was later denominated "the day the Sun brought darkness".

Both these events happened at a time when monitoring solar activity was not all that common. In 1859, it was out of the question even to have satellites, whereas a century later, the Sun was still not under observation 24/7. But things have changed since then. Nowadays, we are stalking the Sun without interruption. We would have a warning if anything of these sizes were to strike.

But besides just observing flares during times of high activity (which might not be on itself a sufficient heads-up), we also have special satellites, designated to record incoming CMEs. They are like space tsunami buoys, as the moment they are affected by space radiation, we know something bad is more or less about to happen. This gives us a few hours preparation time before calamity strikes. Not much, but enough to turn off our devices and safeguard as many components of our infrastructure as possible.

Small storms are enough damage satellites, affect radio communications and prove harmful to astronauts, who are more exposed to it than us. People down on Earth would feel nothing from those as the magnetic field redirects most of the storm and the upper layers of the atmosphere absorb whatever undesirable radiation remains. These miniature events aren't that easily detected, but because of the not so serious effects, no far-reaching warning is needed.

What about the super storms? They are, as I've said, much easier to detect, for we've devised all sorts of mechanisms to help us identify them. But if those mechanisms prove faulty, the buoy probes don't work properly let's say or we don't detect the threatening early sings that early, consequences could be grave. First, high-orbiting geosynchronous satellites would become electrically charged and short out, followed by low-orbiting satellites beginning to encounter increased air resistance as the atmosphere heats up and expands into their flight path. Then, devices on a global scale could stop functioning. Much like in the case of a worldwide cyber-attack, we would be thrown back in the Stone Age for years before society recovers. A severe geomagnetic storm could result in the absolute failure of banking systems, household appliances, servers and mobile devices, and everything else for that matter. This would be bad for the stock market, but beyond that it would establish a regime of anarchy. And trying to recover an entire civilization worth of progress is already hard enough; anarchy would only propel the required time a couple of decades or centuries into the future. Like with the others, this wouldn't exterminate humans directly (it's up to them what happens next), but it would critically wound our world as we know it. Nothing would be the same.

Thinking about it, it is unlikely that we would fail that miserably in the face of a relatively obvious danger. However, it doesn't have to be a full-on miscarriage on our behalf for things to turn out badly. A CME moving faster than normal or an unfortunate series of powerful CMEs might be just more than our preventive measures can take and end up hitting Earth merciless as ever. Also, a single continent thrown into darkness would be more than enough to disrupt society, but I guess that couldn't be considered a true apocalypse. Still, any like event that catches us on the wrong foot could potentially be a doomsday scenario for a large number of people.

2012 – the year the world would supposedly end; we know that wasn't the case. But there were a number of warnings on possibly devastating solar event like the ones discussed in this chapter. In the middle of the summer, on July 23rd, an unexpectedly large and strong CME missed us by a small margin: only 9 days! Even with the warnings, this phenomenon wasn't given that much attention in the real sense. Instead, some interpreted it as a chance of Earth being straight up carbonized by the incoming freight train. The end of the world was in vogue, what would you expect?

Rumors began spreading as to how this is the one true apocalypse that is coming in 2012, further fueled by the mere couple of days' notice. In spite of the spontaneity of this scenario – like all other 2012 doomsday predictions – it didn't happen. But the idea was intriguing. So much so, that many previous and future movies and works of fiction were focused on this "the Sun will burn us alive" type of scenario.

Either CMEs or solar flares, they were all about irregular solar activity of incredible proportions that wiped out life on Earth. Technically, this would be possible if a fabulously powerful CME would hit Earth just the right way, such as to deform our protective magnetic field in an absurd way and then to allow for the highly energetic particles to pass through our second protective layer (the atmosphere) and down to the surface where it could finally burn whatever it wants?! It almost sounds so crazy that it's unbelievable. While we don't know for sure how powerful a CME could get, this scenario is just that – crazy. As for solar flares, there's not much to be said. They happen in the solar corona, not far from the solar surface. And generally, they are of the order of hundreds of thousands of kilometers in size, with the absolute biggest reaching say a million kilometers, but not much more than that. Sounds big, though it's still but a tiny fraction of the Sun-Earth distance. Pretty safe to say there's no threat in those either.

The danger of this rather predictable phenomenon of unpredictable gravity wouldn't be straight-forward. We wouldn't burn or anything. It would be direct in the sense that technology as a near whole may fail, but we would most probably worsen the situation being overtaken by anarchy. The ensuing apocalypse would be far more complex. It is all really intriguing as it is, but we might simply be lucky and get over this sort of scenario in the years to come, just until we are prepared enough to avoid it. We might just have the upper hand on our diurnal orb. But not so fast...

Going down with the ship

As a symbol of life, the Sun may someday symbolize death. Just like any other component of our great universe, the Sun too will eventually decay. It will slowly, but surely, burn all of its nuclear fuel that is currently sustaining nuclear fusion, and therefore powering the beast.

Once hydrogen runs out, it will use up its helium, then heavier elements, all the way to iron, which is the heaviest that can be produced inside stars. The hydrogen era – the Sun as we know it – will last the longest, then helium will run out a bit quicker and so on, as material is lost in the form of released energy. In the very end, the not-massive-enough Sun will not undergo a supernova, so that's not to be feared. Instead, it will take the form of a tiny white dwarf, similar in size to Earth, but with half of the Sun's mass. That would be a very dense body, with rather low temperature (compared to today's), unable to sustain life. Not on Earth, nor anywhere else in the Solar System. But that isn't what we should be afraid of either.

On its way down into the darkness of space, this sinking ship that is our Sun will take the Solar System down with it. It doesn't need to reach the rock bottom to "drown" us all. In about 5 billion years, the stars main fuel source, hydrogen, will be running low. The core fusion of hydrogen will stop, leaving nothing to prevent the core from contracting. The release of gravitational potential energy will then cause the luminosity of the Sun to increase as it continues expanding for the next billion years. The hydrogen remaining in the outer shells of the now giant Sun will start to fuse at this point, increasing its luminosity by a factor of 1000. Because of that, the Sun will gradually lose up to a third of its mass, propelling Earth to a more distant orbit. On the other hand, the two inner planets will be swallowed by the flaming red giant.

When the hydrogen does (for real now) run out completely, a violent phenomenon called the helium flash[2] will occur. A decent chunk of the core will be converted to carbon right away, and the whole thing will shrink to 10 times the current size. Then, much like it happened with the hydrogen, helium will keep on reacting inside the Sun as it again expands for the following 100 million years. The cycle will repeat many times (maybe this is the cyclicity the Sun represents), with mass losses and rapid thermal pulses. Those, too, will be increasing in size and strength as the Sun progresses towards its inevitable death. The Earth's expanded orbit will dwindle due to tidal forces, so, finally, our planet could be completely consumed by the Sun.

After this series of events, only a white dwarf surrounded by a wonderful planetary nebula will be there to remind the universe of how great our Solar Empire once were... The Sun will have reached its end, and so will the Solar System. But hopefully not us!

It is as clear as day that life as we know it wouldn't be able to thrive on Earth when it's literally enveloped by the Sun. Neither would it in the first giant phase. Really, it could be as little as one billion years or even half that time until conditions on Earth wouldn't allow for organic life. So, do we accept that a billion years from now, humans will no longer be?

Without any outside intervention, the lifecycle of a star is predetermined by the universe. It is born, it goes through these weird phases, and then faints until its eventual death. If we were to just carry on with our normal lives on our little planet and somehow not drain all of its resources or sabotage our existence, then yes – a giant oven would probably await to fry each and every one of us (more like our descendants). But this is even less likely than a giant CME actually burning us alive.

Hopefully we don't sabotage ourselves and don't go extinct by then. Whatever happens in the meantime, good or bad, if we don't suffer some other cataclysmic scenario presented in this book, then a billion years is more than enough to evolve, as a civilization, beyond recognition. We would for sure exhaust our planet's resources by then and we'll be forced to emigrate to other places in the universe. In a billion years' time, we might also drain everything there is to drain from our entire Solar System and simply leave it behind. We would, most probably, expand into the depths of cosmos, opening new horizons of possibilities and developing our technology to the very max. Not the Sun, not a single star would pose any issue to us, as by then techniques of regulating celestial bodies and effectively changing their behavior would become a reality. This is the type of external intervention I told you about.

Conclusion

So, provided that we adapt to the everchanging conditions and overcome the obstacles that may get in the way, it begins to dawn that a solar death apocalypse would never come. Almost equally unlikely is for some sort of unthinkable solar flare or CME to snipe us at enormous power in the upcoming period, so as to make us crunchy. Still, the more common scenario, the one presented first, could seriously impede our development, making up for a decent apocalypse as a collapsing society. But solar activity is so volatile and happens so often that we wouldn't normally think an apocalypse would commence this way, which is why I feel the need to explain why I decided to classify it so high up the scenario scale.

It is first and foremost the element of surprise. It might not be completely uncertain that something is about to strike, as we can expect it around times of solar activity peaks. But we are still taken aback whenever something actually happens, given the short notice – from hours to days at most. We might not even expect it in the first place, like it happened in 2012 when it wasn't the middle of a cycle (that came only 2 years later), yet a giant CME was very close to provoking a geomagnetic storm.

Next, we have no reliable defense against a super storm as of now. At best, we make the most of warning systems high up in space that alert us as soon as something triggers them. Basically, our only real defense is acting quickly. If that doesn't happen from whatever reason, be it mass hysteria or a failure to bring the issue to people's attention immediately, a large or at least a reduced-scale mayhem will install.

The severity of a super storm is also though to estimate. While the Carrington event was the largest ever recorded and it did almost have a planetary impact, who knows for sure how gruesome similar events could get? The likelihood of something massive striking us in the upcoming centuries is quite considerable; how massive is uncertain...

I've also come to appreciate the interconnectivity of it all – how many implications a singular unpredictable event could have is just bewildering to me. Others like a global cyber attack might have similar outcomes, but something like that takes much more to "implement". With solar super storms, it's just one little notch that has all it takes to permanently disturb our society. In some ways, this is more severe, too. With cyber attacks you have a number of experienced people working together to remediate the chaos, and it wouldn't take extremely long until things would settle. In this scenario, despite people's collective attempts to rebuild the society, you cannot fight a toasted circuit like you can a corrupted computer. Oh, and don't forget the fact that close to all circuits will be toasted!

There aren't many biblical references to this doomsday scenario. At most you see rains of fire or scorching heat imperiling Earth and its people, and there evidently isn't anything mentioned about electronics failing. Heavenly lights are sometimes brought up, maybe hinting the appearance of auroras, but we shouldn't get too ahead of ourselves. A dark or enlarged Sun may also point at its death and the ultimate demise of our planet, yet that might be inflated as well.

Still, even without a strong symbolic support from the biblical apocalypse, with all in mind, I believe that the solar activity scenario does deserve an important spot on our list. But here is where the real monsters come in...

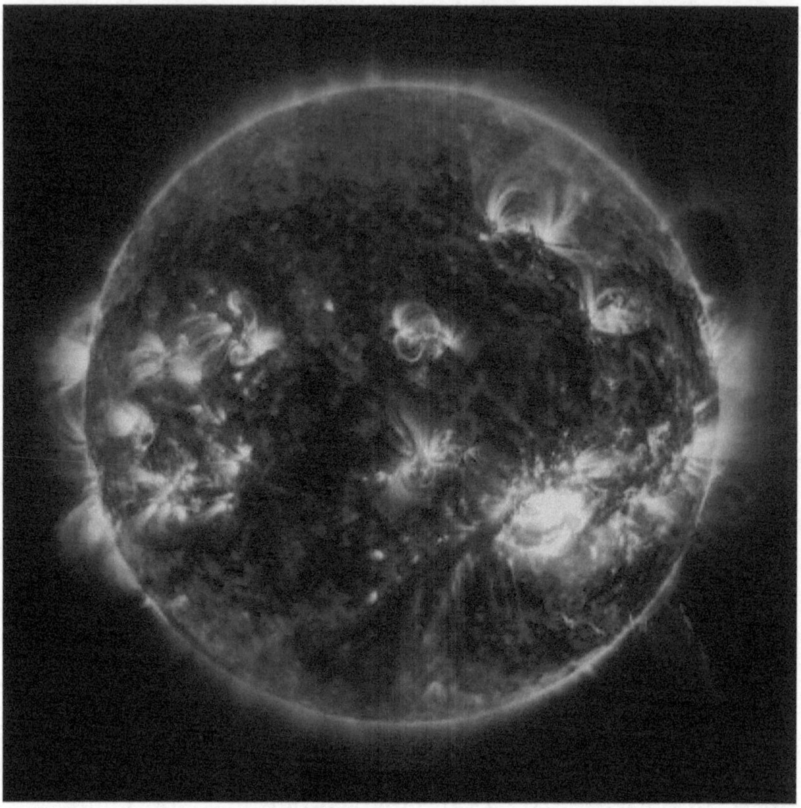

A filtered image of the Sun – it is not as quiet as imagined. You can see traces of hot plasma in the shape of magnetic field lines coming out of the more active regions. Sometimes, those disconnect, giving rise to solar flares, like the one observed in the lower right corner of the image.

1. At high temperatures, electrons orbiting the nucleus get excited and break free from the atom. As the total charge of the remaining atom is now positive (due to the positive protons outnumbering the remaining electrons), it is called an ion. Ions can also be negative, but we are only interested in these positive ones which form the stellar plasma.
2. The helium flash is a very brief period in some less massive stars' lives, in which a thermal runaway nuclear fusion of large quantities of helium takes place. That helium is rapidly fused into carbon in the core, as explained in the case of our Sun. The helium flash increases the density and temperature at the core back again, allowing for further helium fusion.

SECTION IV:

Unpredictable and somewhat likely scenarios

About the serious threats to our survival as a species; events that could happen all of a sudden and against which we have little to no defense

IV.1: Super-volcano

"Volcanoes are one way Earth gives birth to itself."

Robert Gross

Things don't look too good out there, do they? We gain a sense of thankfulness for what we have and how we're doing down on Earth. We live in a moderately advanced society in which the only dangers we face are the unending bills or the fear that you may be fired for speaking up that day two years ago. People have all sorts of problems, that isn't something new. But I bet they were glad of how they're doing at least once, thinking that it could've been worse – what if everything went bad all of a sudden, what then? There might be an answer to that and, this time, it lies just below their feet...

Not even our planet is that quiet of a place. Apart from its thin surface we happen to happily live on, the structure of Earth is more shivering. Though not because it's cold, but because it's extremely hot. An incandescent core surrounded by layers of molten rock and metal. And there is one particular sheath separating the very interior from the outer crust: the asthenosphere.

If you recall, it is the same layer that influenced tectonic movements of the surface plates. What distinguishes the asthenosphere from our planet's other components is that it represents the barrier between the molten mantle and the solid lithosphere just above – the melting point of rock is reached here. It is mostly solid like the layers above, however a slight amount of melting contributes to its mechanical weakness.

In certain regions, the pressure difference allows for additional melting. This is where the asthenosphere wells upwards, shaping like funnels, representing the most important source of magma and creating volcanoes in the process. While this generally takes place at tectonic boundaries (i.e. where tectonic plates meet, especially larger ones), it is also possible that internal irregularities give way to volcanism away from plate boundaries in so-called "hotspots", an adequate name if you'd ask me.

Nothing frightening until now, but keep in mind that the situation isn't always as pink. Consider an earthquake, or an abrupt tectonic movement. Essentially, any perturbation in the hidden chamber of a volcano (where all the gas-rich magma is stored) could trigger a chain reaction, culminating in an expulsion of quantities of volcanic material. This would be a volcanic eruption. Pretty basic as volcanoes go off all the time, just like earthquakes and tsunamis. It's not as if they went off all at once to end the human civilization, though that would be an interesting scenario. But they don't have to in order for that to result. That's frightening. Hold onto something as I tell you that only one of them could cause apocalyptic devastation!

Scaling ticking bombs

Clearly, not any volcano can doom our species on a casual eruption. Not even the bigger ones have such a potential. But when those big ones have a rough day, it is an actual possibility that the biblical stories of fire raining down on us and of a world encompassed in flames come true. From now on, we'll refer to those "fathers of volcanoes" as super-volcanoes.

But let's take a step back. What makes a volcano a super-volcano? It isn't the physical size if that's what you thought. In fact, one may not even be able to determine that. Volcanoes are just like snowflakes, they come in all different shapes and sizes. If the usual description of a conical mountain, spewing lava and poisonous gases from a crater at its summit were true, classifying them would be easier, but that's not the case. For that reason, volcanoes we classified based on their strength, their power. Well, not really... it is actually the amount of ejected material that tells them apart on the Volcanic Eruption Index scale (VEI).

Another similarity with earthquakes (you'll keep seeing the link between the two throughout this chapter) is that the VEI scale is also logarithmic, meaning that a volcano needs to be 10 times "stronger" than another to place one step ahead. But, if a full step isn't reached, then it's still considered the same integer level – fractional numbers aren't used here to highlight small differences. Minor things. Now it's time for the greater.

Among other material, like gases and lava, volcanic ash (correctly referred to as tephra) comes in enormous quantities. Cubic kilometers that is. Of course, smaller eruptions don't make it very far in that regard, with only VEI 5s and above expelling significant amounts of ash. But when you go up the ladder, you get terrified by what could be reached.

VEI 8 is thought to be the highest level of this scale, but not necessarily the limit. Those are the super-volcanoes! They emit over a thousand cubic kilometers of ash upon eruption and have the most serious effects on their surroundings which, at this level, could be the entire planet. Because of the difficulty of accurately classifying a volcano, especially in the lack of previous evidence, it isn't sure how many super-volcanoes are there, but numbers range from 10 to 20 worldwide. When talking about super-volcanoes, you're not to worry whether you're right next to one or a bit farther away. Whatever the case, the aftermath of a hypothetical eruption would largely be the same.

The eruption mechanics significantly vary from one volcano to another, and not all eruptions of the same volcano manifest the same. Actually, there have been hundreds of super-volcano eruptions since the last catastrophic one. But most of those spewed out only steam and gases or small amounts of lava. Yet cataclysmic eruptions do happen and they are completely different than anything of the kind.

Ashes to ashes

What does this mean? Aren't they volcanoes like all others, only more powerful? Well, yes. But the danger is far greater. Not because of the lava that pours down, burning fields and damaging infrastructure. This would surely be an issue, but nothing special to more common eruptions. It would be the ash and the gases that might render an entire continent uninhabitable and ruin society at a global level.

That's not to say that the eruption itself would have unnoticeable direct impact. Nowhere near that. A super-volcanic eruption wouldn't necessarily blow up the entire volcanic apparatus, but it would go off accompanied by an explosion. A decent area surrounding the volcano, a couple hundreds of kilometers in radius, would transform into an earthly hell in a matter of minutes. Flaming rocks would fall from the sky like meteors. Only you would probably not see them because of all the ash and toxic material in the air. This would be so thick that it could cover the entire area in a few meters of ash. Assuming you wouldn't just sit there and try to run, your odds wouldn't be great either as inhaling the unpleasant air would cause the ash to solidify (form a cement-like mixture) in your lungs, eventually leading to death via intoxication. There would be virtually no chance of survival for one that happened to stand there at the moment of the eruption. You see that lava flowing in the vicinity of ground zero would be the least of the problems. So here's where ash and gases make their presence felt.

Those two find it much easier to travel and spread, as they are not constrained by surface boundaries like lava and have the wind on their side. Even thousands of kilometers away, the ash might form some centimeters thick layer that, prior to settling at the surface, would roam in the atmosphere. This would obstruct the Sun, preventing light from reaching extensive areas of our planet. Global temperatures would plummet as a result. Following this drastic drop, plants and crops would rapidly die out, unable of undergoing photosynthesis. Animals would also face massive extinctions as they are hindered from procuring food. I will not even mention the societal and economic repercussions of this, but you get the idea. And the lingering ash clouds would persist for a decade or so. A decade of horror in which we would have to fight for our every meal and for the reconstruction of our wounded society.

And this is not all. The gases would be a whole other deal. In the eruption, gases like hydrochloric acid, hydrogen fluoride, hydrogen sulfide and sulfur dioxide, previously deposited in the underground storages would finally see the sunlight (or whatever traces of it are left). Those would manifest across the continent and even overseas in some form of acid rain and noxious air.

But we shall take a concrete example to see exactly what is going on, let's say Yellowstone super-volcano in Wyoming, USA. The last eruptions took place 2.1 million, 1.2 million and 640,000 years ago respectively, with the first being the strongest of three. Each time, it dealt catastrophic damage to the American continent and drove terrible changes at a planetary scale. Smaller eruptions occurred more recently (known as "relatively non-explosive"), some of the order of VEI 6 or 7, but none to rival the actual super-volcanic catastrophes that marked the distant past.

While looking at past events may not reward us with insightful clues – there were luckily no super-volcanic eruptions since prehistoric times – we could estimate to some extent what such an event might bring about. Assuming it blew up today, tens of millions of people within a thousand of kilometers from the eruption would almost certainly die, with many others to come. The entire United States and Canada would be covered in a thick blanket of ash. Buildings would therefore collapse under this volcanic material as just a couple of centimeters would be enough to cave roofs in. A good illustration of something like this happening would be the buried (now excavated) cities of the Minoans. A VEI 6 volcano in the middle of the Greek island currently known as Santorini covered the entirety of land surroundings in meters of volcanic ash, preserving the Minoans inheritance like a protective blanket of sorts. An even more renowned example is the city of Pompeii and the neighboring settlements buried under the volcanic material ejected in Mount Vesuvius' eruption – strength-wise, much smaller than the others, though.

The ash would also darken the skies as far out as Europe or South America and farther, building up to a nearly 10-degree global temperature decrease which might last even a decade. Don't believe me? Take Mount Tambora eruption from 1815. That wasn't even a VEI 8 eruption, only a level 7. Still, it managed to throw the entire world's population into hunger, dropping temperatures by as much as one degree, disrupting seasonal processes and weather patterns and also destroying vast portions of crops. The following year, 1816, was regarded as "the year without a summer", a name which still reigns true today, when the disturbances can be correctly interpreted. It is considered the deadliest eruption in Modern History (at least 1,300 years, experts say), with its death toll hovering at around 80,000.

Now go a level higher to a potential eruption around 50 times more powerful than that. In a world with today's needs and conditions, the UN evaluated this scenario as so devastating, that the entire world would run out of food in just over 2 months succeeding an eruption. You may want to reflect on the consequences of that for a second.

But there might just be more... Another North-American volcanic monster is situated on the territory of Utah. It goes by the name of Wah Wah Springs and, despite the funny name, it might have been the source of the second most energetic event on Earth, after the dinosaur-killer asteroid. It is though to have emitted almost 6,000 cubic kilometers of ash and covered thousands upon thousands of square kilometers around it in pyroclastic flow. Other eruptions from 130 million years ago might have been even more prolific in terms of ash, some reaching 8,000 to 9,000 cubic kilometers and nearly breaking the VEI scale. But there is one true "mother of all eruptions" that might have triggered, too, on North-American land. It is unsure whether it was a single continuous eruption or a complex of super-volcanoes blowing up at around the same time, but, if the former is true, then in all probability it dwarfed the "casual" super-volcanic eruptions and passed a theoretical level 9 on the VEI scale.

There are many others around the globe: Long Valley in California, Toba in Indonesia, Taupo in New Zeeland, the Central Volcanic Zone in the Chilean Andes, the list goes on. Some might be unknown to us as of yet since they haven't awoken. We can't be sure in this field either – some super-volcano may spring from nowhere and lead to the most terrible eruption in history; maybe some dormant volcano suddenly wakes up and decides to take a midnight snack worth a billion lives before going back to sleep. The one certainty is the ever-pending doom.

440 million years ago, the Ordovician extinction came up. In a time when marine life was predominant on Earth, super-volcanic eruptions are believed to have had an incredible importance in driving this event. The newly created volcanic rock would have been eroded by water and wind and consequently reacted with the atmospheric carbon dioxide and absorbed it. With this radical reduction of greenhouse gases, temperature dropped substantially, water froze and, after a couple of these cycles, 86% of the existent species were lost forever. Moreover, the Permian-Triassic extinction – the biggest mass extinction in Earth's history – may have also come about thanks to a super-volcano. Billions of tons of volcanic gases destroyed the ozone layer, the average temperature of oceans exceeded 40 degrees Celsius, acid rain fell all over and the greenhouse effect was unbearable. While the title of world destroyer is fought between that and an asteroid impact (which might have led to that), the 95% of ocean species and 75% of land species are long gone.

So, if we know that danger is awaiting, then we have definitely devised some kind of prevention or prediction mechanism at the very least, right? What would you expect? Of course we don't! Otherwise super-volcanoes wouldn't be topping our lists. But let's not panic and just righteously assess the situation.

Against all odds

Firstly, should we worry? Well, yes and no. Yes, because there hasn't yet been any super-volcanic eruption in human history. The last ones were tens of thousands of years ago and we know they didn't go very well. While humans were few back then and seemed to have not gone extinct, you can bet they suffered from colder weather and insufficient food. Lake Toba erupted 70,000 years ago and released 3,000 cubic kilometers of ash. The most recent eruption was 25,600 years ago when Taupo went off in New Zealand and had similar consequences. So, yes, we should be worried because of the frequency of those explosions. You see, it isn't just Yellowstone which erupts roughly once every six or eight hundred thousand years. There are many doomsday-bringer candidates on our planet; hence, the eruption frequency drops to around 50,000 years. This is also why we shouldn't fear this type of apocalypse according to some – they are quite rare. It is relatively unlikely that one would erupt in the next 10,000 years or so, which would be more than enough time for us to develop and overcome this sort of mundane threats.

Still, there is this unpredictability feature of theirs which makes up the second point. There is no definitive rule which applies to super-volcanoes. Not regarding their shape, not their volcanic material output and especially not the frequency of catastrophes. Given the previous eruptions, there is no way to clearly provide an approximation of when the next one will happen. But if not for thousands of years of "heads up", maybe a decade or less would be reasonable?

Scientists say it is extremely challenging to try and predict when a super-volcano might erupt again, but that didn't stop them from diving deeper into the issue. Analyses revealed no single, unified model that described how each of the events played out and showed that the super-eruptions could start mildly over weeks to months or go into vigorous activity immediately. The individual super-eruptions themselves could occupy periods of days to weeks, or be prolonged over decades.

Teams of researchers investigated the potential early warning signs of such a world-shaking eruption, focusing on the volcanoes beneath Lake Toba and Lake Taupo. Evidence from the Youngest Toba Tuff suggest that the eruption began abruptly, with the immediate collapse of the chamber roof. In contrast, the Oruanui eruption (Lake Taupo) started slowly, depositing a large ash blanket before the caldera collapse, and progressed intermittently including pauses of several months. "The uncertainty associated with these events therefore makes it very challenging to determine when and how these volcanos may potentially erupt in the future." Said the experts. Moreover, the build-up may be slow and silent, leaving us with no hints whatsoever. We are left clueless before some of the mightiest apocalypse heralds... But scientists attempt to calm us down by restating the low probability that a super-volcano will disastrously erupt in the upcoming centuries and millennia.

A mind-twisting paradox is relatable to this apocalyptic scenario: The unexpected hanging paradox revolves around a prisoner on death row and the judge which sentenced him. At the end of the trial, the judge told the inmate that he would be executed sometime next week by hanging[1], but the day he will be executed would be a surprise. Troubled by the unclear character of the sentence, the prisoner comes up with an interesting reasoning. "If the execution was set for Friday, then I would be alive by Thursday, and I would know that it was going to happen on Friday, therefore it wouldn't come as a surprise." – he ruled out Friday. Moving on, "if Friday isn't an option, then Thursday would be the last possible execution date, but that too can be ruled out as it wouldn't be a surprise since I would be alive on Wednesday" – there went Thursday. He continued ruling out each of the week's days, concluding that he can't be executed anytime this week. He was in cloud nine for a while, but then on Wednesday, when the guards brought him in to the gallows, he was completely surprised!

We can keep saying either that it is likely that it will end us all, or that the mass hysteria is foolish, but a super-volcano apocalypse would be very surprising indeed, just like the inmate's hanging. Another chapter, another grisly acknowledgement – I know. But perhaps there is a solution to prevent one from happening, if none to predict.

One particular plan that scientists put together at NASA JPL 2015 stands out. They primarily focused on the Yellowstone volcano, considered to be the biggest threat on the American continent, but the ideas are somewhat "applicable" to other super-volcanoes as well (we'll see in a bit why "applicable" might not be appropriate in the context). They proposed that we cool the volcano down.

The plan was to drill a series of wells around the perimeter of the volcano, some of the deepest in the world, more than 10 km below the surface. Then, cold water was supposed to be pumped in the wells and over time cool the chamber. The project also involved the water be pumped back up to be used as a source of energy after it was heated in the chamber.

The proposal was ingenious or, better put, inventive. It would make use of available materials and terrain and it would also provide energy from the process. As far as efficiency goes... it doesn't go very far really. A sustained 20 Gigawatts of power would be required to cool the volcano down to a certain, practical temperature, and the entire operation would take about 16,000 years to complete. The volcano might naturally erupt before we achieve our goal. The estimated 3.5 billion dollars this initiative would demand isn't that big of a struggle, but the sum still drew some attention. But let's say we would put all the effort into it so that its finished in far less time. What went wrong that the project was rejected?

The risk factor was a worry to many. Cooling the rock could create fractures near the magma chamber, which might spontaneously trigger a super-eruption, the very thing we are trying to avoid. All in all, it wasn't worth attempting – so, applicable? not so much.

Conclusion

So we are left with the following: no precise prediction method, only speculations, not a single applicable prevention mechanism, just ludicrous plans, and a constant super-volcanic threat to our pure existence. An old Romanian saying went like: "the sword won't cut the humble head", and it was often contrasted with a supposed Turkish one that said the exact opposite: that the humble head is meant to be cut. Which do you think works best in this context?

Proverbs or not, this volcanic swordsman has a short temper (of the order of thousands of years, but still short in the geologic timeline). And his sword is sharp! Jokes aside, we are under a sword of Damocles, unaware of when it may fall, and currently unable to stand up against it.

People have always manifested a fear of volcanoes, and the reasons haven't changed throughout – we are powerless in the face of them. Yet for the most part, they didn't even grasp the greater danger. No one knew about super-volcanoes until recently. And if one were to go off now, the reality would turn out way darker than any historical depiction of volcanic eruptions. Or maybe not... maybe the very Apocalypse portrayed in the Holy Books is a closer comparison.

"Then the angel took the censer, filled it with fire from the altar, and hurled it on the earth; and there came peals of thunder, rumblings, flashes of lightning and an earthquake. [...] The first angel sounded his trumpet, and there came hail and fire mixed with blood, and it was hurled down on the earth. A third of the earth was burned up, a third of the trees were burned up, and all the green grass was burned up. The second angel sounded his trumpet, and something like a huge mountain, all ablaze, was thrown into the sea. A third of the sea turned into blood, a third of the living creatures in the sea died, and a third of the ships were destroyed."

Revelation 8:5, 7-9

"When he opened the Abyss, smoke rose from it like the smoke from a gigantic furnace. The sun and sky were darkened by the smoke from the Abyss. And out of the smoke locusts came down on the earth and were given power like that of

scorpions of the earth."

Revelation 9:2-3

"The horses and riders I saw in my vision looked like this: Their breastplates were fiery red, dark blue, and yellow as sulfur. The heads of the horses resembled the heads of lions, and out of their mouths

came fire, smoke and sulfur. A third of mankind was killed by the three plagues of fire, smoke and sulfur that came out of their mouths."

Revelation 9:17-18

The more you inspect those readings, the greater the similarities become. It is quite chilling how scenarios like this might have been foreseen that long ago, in the lack of extensive knowledge, and that they also sit at the foundation of Christianity. I am left speechless of how to cheer you up a bit, other than restating that it's relatively unlikely that something like this would come upon us in the near future. But that's still relative. Better said: there are a few other ways the end of the world could come which are, not exaggerating, much more probable than this one...

"An asteroid or a super-volcano could certainly destroy us, but we also face risks the dinosaurs never saw: an engineered virus, nuclear war, inadvertent creation of a micro black hole, or some as-yet-unknown technology could spell the end of us."

Elon Musk

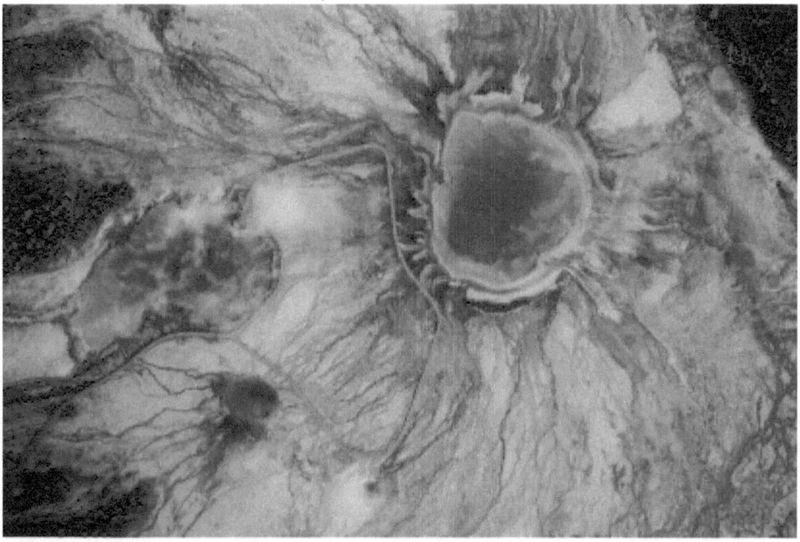

A satellite view of Yellowstone National Park, a nature's wonder under which lies a dormant super-volcanic beast

1. Most often, executions do not happen on weekends, that's why the inmate begins his reasoning starting from Friday and going backwards.

IV.2: Global pandemic

"At your highest moment, be careful - that's when the devil comes for you"

Denzel Washington

Words of wisdom coming from the mouth of a world-renowned actor, which also has an accentuated inclination towards Christianism. And I couldn't help but notice how these words resonate incredibly well with today's post-pandemic society.

We are living in a world very different than a hundred years ago. I'm not even going to mention how utterly distinct today's world is from prehistoric times. But upon reflecting on the mass hysteria that this last global pandemic sparked, I've come to the realization that our world is very much unlike what it was right before that calamity stroke.

In just a bit over two years, our world has changed in unprecedented ways. Our interactions, our lifestyle and, most importantly, our mindset are no longer the same. Before the infamous SARS-CoV-2 virus appeared in the titles of almost every publication, I, for one, had no actual idea how a deadly disease (that's what it was after all) could make the flesh of every human being creep – not in the proper sense, but still.

I've heard of the Black Death and the Spanish Flu which decimated millions and millions in earlier times. In fact, I remember that I was just reading an article about the impact smallpox had when I heard the terrible news. Another virus began striking fear and death among the Chinese, and sporadic cases had already appeared in multiple nations around the globe. I couldn't believe my eyes at first, but soon realized that an "east wind" was coming, and it was coming hard.

Also, it's fair to say that I have highly underestimated the potential of such a virus. I was not prepared. One day we were talking about how the first cases appeared in Romania and the next day every event that I've planned attending to was being cancelled. Less than two weeks, that's how much it took for all students to go into isolation. In less than two weeks, my whole life was seen from a totally different perspective. Classes were halted and I wasn't allowed to leave the house, let alone see my grandparents who were thought to be the most sensitive to the disease. All of a sudden, daily disinfections of remote controls, keyboards, doorknobs and what not were established at our place and hand sanitizers became everyone's best buddy. I took things to such an extreme that for over a year and a half I've been opening doors with my elbows only. My parents who were still working for a while longer never took off their face masks and family dinners were no more. Online meetings, which I had previously never heard of, spontaneously became the main means of social interaction.

But the airborne predator still got me and my family a couple of months in the pandemic. I was confined to my room and proximal bathroom for two weeks straight, not touching my face and not breathing in the hallway even though I was aware that everyone had got the virus anyway. It was all a fearsome experience which I wouldn't have bothered imagine had I not lived through it myself. I was among the lucky ones who made it through unscarred. It really is a game of dice once you get to the critical point of infection. So it is with all viruses and diseases, but especially with those that don't have a certain cure[1].

Countless frights that one of us might have got the virus again later, spanning across over two years of social distancing and extreme precautions, the game of dice played out in the favor of humanity. But COVID-19 was no easy feat, with a confirmed half a billion cases and more than 6 million fellow earthlings losing the fight against it. Yet maybe, after all this suffering, our civilization has undergone a shift in perspective regarding these unseen enemies. I highly doubt that's just me perceiving things differently now, but I am still unsure whether this is applicable to the whole of the human race. This is what scares me – the fact that even in this modern post-tragedy world (some extremists go as far as to say "post-apocalyptic"), we haven't learnt our lesson.

If there's something we were truly not ready for when we were supposed to be, it's got to be this pandemic. The Black Death and the Spanish Flu were much greater in size, but their consequences were understandable for the given social context in which they ensued. For COVID however, there is no reasonable justification for how in the world did it happen that, in an extremely developed world compared to what was back then and with so-thought "exceptional" medical conditions and information techniques, we only managed to escape this curse two and a half years in out of pure luck.

We should be glad that it's finally over, but we must not settle with that, for something much worse could potentially happen. Could COVID be a mere harbinger of a far greater adversary? And how could the situation take such a rough turn for humanity that it might not be able to recover?

Handling an outbreak

We cannot control viruses and diseases, that's why they are called outbreaks. Whenever one happens, it won't stop until it reaches an ultimate barrier: either it dies out, or people gain immunity against it. Both would work, optimal measures need to be taken in order to protect ourselves efficiently, meaning that we wouldn't want to let millions die just so that the virus dies with them. So, how do we go about it?

These peculiar illnesses spread much easier than your normal cold. As in most cases, a random person sneezing might suddenly infect 10 more nearby. Either by direct contact, or by food, water and air transmission, any feeble act may lead to new infections. This is how epidemics are born.

The main difference between an epidemic and a pandemic is that the former refers to a large number of people infected, but within a certain region or community, whereas the latter takes it to a much larger scale, that of multiple continents and, virtually the entire world. Actually, it all starts with a few endemic cases, restricted to a particular people or country, but when saying "outbreak", we understand an unexpected increase in those cases, leading to an epidemic and, later on, to a pandemic.

Now that this is clarified, we are still left with a highly contagious disease (i.e. very infectious). If distinguishable symptoms appear right away, relatively soon after one is infected, then we might have a clue that we are not confronting the usual colds we get. Depending on the severity of those and on our ability to respond quickly, our best chance is to try and isolate the virus. This might mean quarantining a specific area around ground zero[2] so that no other people get infected.

However, this is generally not as easy as it seems. It is fairly common that one does not present drastic symptoms immediately upon infection and isolating entire communities may simply not occur to us. By the time we acknowledge the unforeseen character of this foe, further infection might just be facilitated enough that it spreads to other countries and regions. At this point, three strategies may be implemented.

The first one resembles a natural selection process, wherein we try to race through the epidemic (or even pandemic) by counting on our ability to develop immunity. We would impose no measures of preventing infection and leave the disease do as it pleases, hoping that we would eventually become resistant against it. But this is a bit absurd. If anything, we are trying to avoid a game of dice, not provoke one. So, we wouldn't risk the lives of millions just so that the pandemic ends faster this way. Yet sometimes we do race through epidemics even unknowingly (without severe symptoms showing up), but that wouldn't be that much of a panic to begin with.

A second solution would work better in terms of health-related risks, but it would leave a significant stain on all the other aspects of society. That would be to mass-coordinate the spread of the virus, so that it would sometime starve to death. With serious restrictions regarding travel, social interactions and any kind of external activity that might clear the way for new infections – that happening simultaneously in all exposed areas, be that local or worldwide – it would be possible to terminate the pandemic. Carried out properly, this could end a pandemic in a matter of months, with low loss of life. But unless the virus is effectively eradicated, which is unlikely, things might escalate back again due to other carriers, such as animals. This strategy too may be a bit far-fetched.

Another roundabout focuses on buying us ever more time. By practically delaying the spread of a virus, we give research facilities time to produce a vaccine. Tactics which may include wide-spread testing, identifying carriers, quarantining exposed individuals and maintaining a general social distancing. Still, the virus will slowly spread, with some communities continuously struggling and others keeping things under control and returning to business as normal, reenacting the measures in case of a resurgence. Delaying will imminently bring a number of people meet their demise, but in months to years, one or several vaccines might just be available to everyone thanks to a joint international effort. Once a decent portion of people get vaccinated, herd immunity kicks in and the pandemic fizzles. This is usually the most viable strategy for newest outbreaks, uniting all these various techniques to lessen the damage as much as possible.

We saw this happen with COVID and we know the story. It began at the end of 2019 and rapidly spread to almost all of the world's nations. In 2 months, it got to Romania and, two weeks later, nation-wide isolation came into effect. All borders were shut down in multiple countries and several other extreme measures were taken. Even so, 6 million isn't a shy number for a death toll. Where did we go wrong?

Our technique was a form of the latter – delay and, later on, vaccinate. But there were two distinctive factors which prolonged this pandemic for two years and a half and constantly mocked researchers and officials. One couldn't regard them as unexpected. We were totally aware of them, but we somehow failed to treat them with proper caution.

We live in an interconnected and increasingly globalized world. People travel all over the place 24/7. Only airline companies estimate some 100,000 flights daily. Add that to the number of road and naval transportations and you get an idea how easy a virus finds it spreading. Back when humans were primal hunter-gatherers, they were always on the move, never staying in a place long enough to cause mass contagion. Viruses and microbial infections were contacted much easier once agriculture was invented and people started to live in communities, alongside cattle, which were also often times disease carriers. But now, things are escalating much quicker. Perhaps the measures were not drastic enough. At the same time, misinformation and non-believers played their part in spreading the virus further. Either way, the virus was extremely infectious and that was reflected in the half-billion people who got it. The consequences were understandable given the course of action we opted for, but something else had even less mercy on us.

Apart from its contagion and moderately high virulence[3], SARS-CoV-2 harbored one trait we had much trouble dealing with. Pertaining to viruses and some other sorts of pathogens, there is a hidden perk in combating them – managing mutations. Expected, in the sense that we knew mutations were meant to occur from time to time, but unexpected as to how often those would occur and how little they would resemble the older variants. As time went by, the variants differed in infectivity and severity, but also in structure. That made some harder to counteract with the already existing vaccines, rendering those ineffective against novel mutations. What I meant when I said we were lucky the pandemic ended (for now...) is because the latest variants do not threaten us as much as earlier ones did. The first variants faded away, while new ones are not nearly as severe. To this day, the vaccines are only partly effective against most variants, but a wide-spread immunity may have been achieved. Yet the virus is insatiable and, since it hasn't been eradicated, might very well evolve into a whole new kind of respiratory diseases, for which we do not have a remedy, not even partial.

For this, we weren't ready. As Bill Gates said when comparing pandemics to other natural disasters, "these pandemics only come along so irregularly that being lulled into a sense of security where it probably won't be a problem in the next few years, why should we put money into that? You don't buy the insurance policy, basically.". This was our foremost mistake in facing COVID. Regardless of the massively developed healthcare system that we thrive upon (compared to previous pandemic times), we have yet to master the techniques of devising impactful vaccines at a reasonable pace. In the research and development (R&D) area is the one in which we should put the most dedication. We ought to mature mRNA so that we can increase mass production, while also keeping it cheap, thermal stable and, of course, reliable. Therapeutics and diagnostics also have room for improvement, and field work shouldn't be neglected either.

But those suggestions aren't something new. We could've taken these precautions, we could've built our insurance so to say, long before COVID hit. In fact, Gates held numerous speeches and tried to draw people's attention years in advance. At one particular TED lecture of his from 2015, he claimed that "if anything kills over 10 million people in the next few decades, it's more likely to be a highly infectious virus, rather than a war. [...] We've actually invested very little in a system stopping epidemic. We're not ready for the next epidemic.". And guess what? We weren't ready. It might have not been 10 million, but it wasn't that far. It sure was disappointing seeing people having ignored all warnings when disaster stroke. If not considered seriously, the lack of these insurances will leave us at least equally vulnerable to subsequent outbreaks.

The pre-COVID era

In all this unpreparedness characteristic to us, SARS-CoV-2 was not nearly as devastating as it could've been. We saw the abnormal struggle it caused, but things were a lot more under control than say a hundred years ago. We found it so weird because our generations haven't faced anything like that before. Frankly, our ancestors might have found it just as weird, but this time, not because they haven't faced pandemics at their time, but because of how different our approach is.

Previous pandemics were treated not with indifference, but with a flawed perception of how diseases behave. The Black Death of the XIV century was just as the name implies, grim. Doctors were few and the principles of modern medicine were unbeknownst to them. Few were educated and even those often appealed to fallacious techniques. Most of the time, the disease was not cured, but amplified by misbeliefs of the time. In an attempt to extract the so-thought sources of the plague for example, doctors used methods we would now regard as torture only to spark an additional infection that would kill the patient way before due time.

That was in the Late Middle Ages, when the bacteria called Yersinia Pestis made its way to Europe. Thanks to the rise of commercial routes, diseases could then be spread much smoother than before. It was especially the Silk Road which connected important states of Asia, Africa and Europe, and while the movement of goods could be carried out flexibly, so could the transport of germs. This is how it's thought the bubonic plague was born.

With the Black Death, it wasn't a virus acting up, but a bacterial. It didn't relentlessly undergo mutations like a virus would, but the living conditions of the XIV century facilitated the pandemic in unprecedented ways. Crowded neighborhoods, the accumulation of waste in a disorganized society and the ominous unsanitary conditions were the main actors in providing the plague with the perfect spreading environment. Whenever one person out in the streets sneezed (without a mask), many others were exposed to the disease before anyone knew how it would manifest, before the sneezer even presented symptoms. This way, almost all of Europe was infected with the plague and approximately 50% of Europeans died in just a couple of years. Worldwide, some 35 million people met the terrible fate – that is 10% of the entire population at the time. But do keep in mind that not everyone was as struck as Europe, that is the plague didn't spread all over the globe.

Physicians had those infamous bird masks, with the beak being supposed to protect them from contacting the contaminated air. That didn't work out very well... Per general, the Black Death had a mortality rate of almost 70% among untreated case, with less than 15% of cases being treated. It is thus regarded as the deadliest pandemic in all human history. It's worth mentioning that over the span of 50 years, up to 200 million people perished because of the bubonic plague: topping over 40-50% of the total population!

But that wasn't the first time Yersinia Pestis had a quarrel with our civilization. Thought to have originated in northern Africa, in the spring of 542, a first plague arrived in Constantinople, but rapidly spread across the Eastern Roman Empire (the Byzantine Empire) and the Sasanian Empire. This Plague of Justinian (named after the Byzantine emperor Justinian I) concluded in 25 to 50 million deaths – again, over 10% of the world's population. Plagues such as that of Cyprian in the third century AD also brought Europe to ruin for decades. Other waves have followed since, depending on how the bacteria evolved. Mass graves can be found in cities like London, where the plagues had a particularly strong impact.

Another pandemic of terrible proportions is known as the Spanish Flu, which broke out near the end of WWI. It was caused by the H1N1 Influenza A virus and behaved quite similarly to SARS-CoV-2. It came in four waves and finished off some 50 million, with estimates going as high up as 100 million. That would make it the second deadliest pandemic after the Black Death. An improvement in medical techniques could be seen, but rudimentary measures were also taken. A common habit was to throw all of the infected and their contacts into the same bowl of a mass hospital ward. In retrospective, it did somehow work out and the flu (this strain of it) came to a close. But people were fighting another disease at the same time, the one with the most casualties in history.

We are talking about smallpox – a disease caused by the Variola major virus. Having suffered a myriad of mutations over the years, it killed more than 300 million people since 1900 only. In total, the numbers may even reach 600 million! Unlike the others, this one affected almost every region on Earth, making it extraordinarily pernicious. Thankfully, smallpox is one of the two officially eradicated human diseases as of 1980, as declared by the World Health Organization (WHO). This one we owe to vaccines!

Those have lately become anything but a luxury. For the most part, everyone has access to vaccines for many illnesses. Their impact can be seen in society, as well as that of antibiotics. Regarded by many as the most important discovery ever, they have essentially transformed our lives for the better. The life expectancy significantly increasing in the last decades is a great proof of that. But some are still skeptic about them, trusting their own immunity above certain medication. And they might have their reasons.

Now, don't get me wrong. I am not here to adjudicate anything that's not related to my area of expertise. However, I do consider it necessary to go through some of the queries in order to understand how we might deal with future events of this kind. The masses are always unsure at first of whether remedies can be trusted or not. Just imagine demanding all people wear masks at the time of the Black Death. "Masks? What for?" Their grip on bacteria and viruses was almost non-existent. There was no previous experience in which masks did prove useful and, with the infectivity of that new disease, they might have seemed futile. Interestingly enough, many trusted physicians with their extravagant techniques, even though physicians themselves often had no idea what was going on.

One would be inclined to believing that, in the XXI century, at the apex of medicine, things would somehow change. Or that would have been the case before the pandemic, at least. What we saw was far from expectations. It wasn't just a person or two who thoroughly refused to wear masks, but entire communities. Scandals and protests arose as to how one should be responsible for their own choices, while carelessly putting others at risk. Many were fearless until they got to feel the taste of it themselves. They were only waiting for the cure, and when the cure came, they still weren't happy. The importance of vaccines cannot be overemphasized, but it can be so easily misapprehended as a result of the social context.

The hurry in which vaccines against this virus were haplessly developed was another thing. Attempts which worked out pretty well against certain variants were taken aback by new mutations. And as we have this "gift" of highlighting the bad outcomes over the successful ones, various opinions appeared on the need and benefits to get vaccinated. Adverse reactions were widely feared, too. You could've theoretically presented unjoyful responses upon vaccination, but that wasn't out of the ordinary among other medications. Why would you take a Nurofen pill for fever then? You could technically go into a coma or a cardiac arrest after swallowing it, according to its side effects. It is compulsory that such information is presented to the consumers. But we do take it every now and then, completely ignoring that information – because it happens so rarely! Yes, you do take a risk when vaccinating, just like you do when taking any medication. Everything in life is a risk one way or another. It all comes down to which risk you choose: that of getting a slight fever or arm pain after the vaccine, or that of you and the ones around you ending up in a much worse position because you got the disease.

Apocalyptic diseases

"Through globalization, the world shrinks."

Dahlia Schweitzer, author of "Going viral"

The pre-COVID era of outbreaks does look a lot more gruesome than what we had to deal with. What was panicking about the SARS-CoV-2 was not its severity or mortality rates, those didn't even compare to other diseases throughout history, but the speed at which it engulfed the entire planet. In this busy world of ours, we were disorganized as always, and the virus had such a good time travelling from one place to another.

But pathogens are constantly evolving. They aren't waiting for us to overcome these chaotic foundations our society seems to be built on. And just because this pandemic barely ended (maybe not even that), it doesn't mean that another couldn't spark spontaneously and disturb our recovering from the last one. This is one of those "don't-know-until-it-happens" scenarios: we couldn't estimate the odds of something like this striking today or tomorrow, but what if it did?

"Although the chance of a disaster to planet Earth in a given year may be quite low, it adds up over time and becomes a near certainty in the next thousand or ten thousand years."

Stephen Hawking

With the current technological development and a bit more put into the field of medicine and research, in the future it might be possible to attain a status of relative resistance to most viruses and maybe even predict the possible mutations. Well, perhaps that last part was too exaggerated, but the point is that healthcare is set to significantly develop and reach new heights in the not so distant future. And while plagues might still mess with us later on, it is now that we should be worried about.

No matter how destructive the pandemics got so far, we did manage to avert an apocalypse on every occasion. But with the right conditions and environment, anything is possible.

I remember lying in bed and watching the 2014 FIFA World Cup when something caught my attention. Some TV had the volume turned high up, and I heard the news from the other room. It was some sort of epidemic that killed nearly everyone infected with the virus in a few-days' time. Symptoms were acute and severe and the patients would often bleed to death (that's what shocked me). It was the Ebolavirus outbreak in Central and West Africa – Ebola hemorrhagic fever, that's what the disease was called. I was in terror! I missed the Netherlands playing, my favorite team back then, to listen to the entirety of the news fragment and do some individual research afterwards. My ten-year-old self was certain that the apocalypse was coming.

Looking back, it might have been because of the explicit imagery I came across online, but people were scared. It wasn't nearly as bad as the COVID pandemic, yet the wide-spread panic was present in the beginning days of the new outbreak. And there are a few reasons it didn't reach such a potential. If only I knew them earlier...

The virus was an extremely lethal one. This might seem worrisome at first, but it's a first reason why it didn't reach the "pandemic" title. Because of the high mortality rate and distinct features, people caught on really quickly that they were dealing with some abnormal disease. Dire symptoms also appeared pretty early on relative to the moment of infection, which gave doctors a heads up that the virus could be contained right away. By calculating the worst-case scenario starting from patient zero and observing how people in other areas were behaving, they were able to establish a perimeter around the infected region, and isolate it from the rest. Accurate computer algorithms are lately the best tool for doing that.

The means of transmission worked on our side, too. Fruit bats are believed to be the natural hosts of the Ebola virus, meaning that one could get it either through contact with infected bats, or through intermediate hosts – infected animals. It is then passed on to other humans by contact with bodily fluids. Doesn't sound good, but it isn't airborne like COVID or other viruses, which is in itself advantageous to us.

Lastly, the area in which it originated also played a great role in this outbreak. I'm not talking about bats here (COVID possibly started from bats and look how far it got), but rather the geographical and demographical area where it first appeared. It wasn't in a densely populated region, and transportations that facilitate the spread couldn't be compared to those in crowded and more developed parts of the world. It thus didn't get very far.

It just so happened that the most fatal of the infectious diseases (apart from a special one we'll talk about later) with no viable vaccine was contained and ended at less than 9,000 deaths. Other variants popped up, but a vaccine has since been developed and approved. Ebola was not to be feared. But some of its cousins could get the job right.

What they'd need to do is the exact opposite of Ebola. To begin with, they should stem in a vastly populated area. Major cities like Tokyo, New Delhi, Shanghai, Sao Paolo, or others would do the trick. Actually, Tokyo might not be the best starting point, as everyone would act carefully the second symptoms appear. But anyway, that brings us to the second point: an extended incubation period. In the absence of symptoms, a pathogen could easily spread to many people without anyone noticing. Something like a couple of months would already be terrible. In order to infect as many people as possible, air, water and food transmission would need to be of great proportions. Because intense travel is paramount in major cities, the pathogen could spread to many other territories prior to symptoms showing up. People on the other side of the globe could spark new hotspots just days after the initial cases. Land and air carriers such as rodents, insects, birds and livestock would make it sure that the vast majority will contact the virus stealthy. It would also come in handy that the sanitary conditions of ground zero aren't the best, to put it this way. Shortcomings in healthcare and sanitation systems would only propagate the disease further in the incipient period. Developing drug as well as thermal resistance would hugely benefit the spread, being able to maintain a similar behavior even in extreme conditions. Once a considerable number of infections is reached, the pathogen is free to lay bare symptoms.

Either via mutations or antigenic shifts[4], the agent may manifest abruptly. It could be small things so people would first seek basic medication, and then as many would do the same thing and find out nothing works, more extreme symptoms like tumors, systematic failures, seizures, comas and hemorrhagic shocks are good to go and finish off the infected. Of course, this would be gradual, but it may also come directly as deadly manifestations. However, that would not be of much help to the pandemic, as it would eventually die out with its victims before it would reach its full potential. No matter what, worldwide research for devising a vaccine would promptly be conducted. Once it gets here, sudden alterations in the pathogen's structure could render any such endeavors useless, giving it time to infect the remaining people and devastate our society.

This is how a disease could theoretically take apocalyptic proportions. I wish I could tell you that reality says otherwise, but unlike with other scenarios, this really has chances of happening in real life. It doesn't need to present all of the traits mentioned before, just some of them would be enough to cause extreme damage and be deemed apocalyptic. The main aspects would be its origin and the extensive incubation period (apart from, obviously, eventually proving lethal). After that, in a mixture of mutations and human error, the virus would spread throughout the world. COVID started in Wuhan, an 11 million people city in China, and was somewhat mild in the first part, with about a week-long incubation period. If only it took longer to incubate and it were a bit harsher in terms mutations, things could've got nasty.

Now, it is easiest to talk viruses, as they are the most common form of pandemic disease. Think of Influenza for example, more commonly known as flu. It is so common we don't even regard it as terrible anymore. Many of us have caught it in the past, but it was just like a rougher cold. But over the years, the flu accounted for some 500 million lives, similar to smallpox. It culminated with the Spanish Flu in 1918, yet every 20 to 40 years or so, the virus undergoes a dramatic mutation. But pathogens like bacteria, such as Yersinia Pestis, parasites or others may also inflict calamity upon the human race, only by slightly different means.

An interesting possibility, quite unlike what you would expect from a pandemic, might come from well-known realms. The Zika virus is notorious for being transmitted to infants during pregnancy. This leads to all sorts of birth defects, among which count microencephaly and severe brain malformations. Bad enough as it is, but provided the virus suffers some misfortunate mutations that allow it to survive in various climates (not just in a narrow belt around the equator, as it's described) and gives way to near-total mortality among infants, a very morbid scenario could be reached. Something along the lines of P. D. James' novel "The children of men", in which potency plummeted to zero, and no children were able to be conceived, thus endangering the perpetuation of our species. This, however, is quite outlandish of a scenario, but nothing is sure.

But as I've said, maybe the treat doesn't come from virus diseases. Those are particularly spontaneous with their steady mutations and could come up any time, be it a new string of some perilous virus or one that's been put away from us for a long time. With viruses, everything would be quite unexpected. But what if I told you that some other apocalyptic disease is coming, and that we know for a fact? We know exactly what it is, we just don't have any cure.

Total bacterial resistance

Similar to how genes and chromosomes in eukaryotic cells mix up in order to make up as many gametes as they can, prokaryotes (i.e. bacteria) also go through their own process of recombination. A key difference is that bacteria do not have their gender determined morphologically, as it happens with their counterparts. That is, it's not genes which tell whether a cell is male or female, but some other biochemical components.

Bacteria have chromosomes in their structure, as well. The ones with only the bacterial chromosome are female, noted F^-. Those which, besides the chromosome, also present a double-stranded DNA fragment called a "plasmid" (or fertility factor F) are considered to be male, noted F^+, or Hfr in case they distinctive (standing for high frequency of recombination).

When an F^- female bacterial meets an Hfr male, their conjugation takes place, with a view to transferring DNA from the one to another. Plasmids are physically separate from chromosomal DNA and replicate independently, so the male makes a copy of this fertility factor and gives it to the female, which consequently becomes an F^+ male. However, by accident, when the F factor is copied in the male bacterial, it is possible that a little fragment of the chromosome to which it's attached is also copied in the process. Therefore, the previously-female bacterial accumulates additional genes from the male.

What's the problem with that? You see, many of those genes give the organisms resistance to various antibiotics – substances that act against the bacteria. So, let's say that the female has a gene which confers resistance to erythromycin, and the Hfr male has, among others, a gene for resistance to ampicillin which is accidentally copied in the process of recombination. Upon that happening, the first bacterial acquires the ampicillin-resistant gene from the second, thus becoming immune to both antibiotics. This way, through repeated conjugations, bacteria become more and more immune (resistant to more and more antibiotics). It means that, given enough time, bacteria will intrinsically attain total resistance.

As of now, there are a bit over 100 antibiotics in the world. Already, one of three antibiotics prescribed is, on average, inefficient against the assigned bacterial infection. The thing is that there are few antibiotics out there, and we'll have an ever-harder time identifying new ones. Besides, as bacteria multiply once every 20 minutes or so and some recombine at around the same pace, despite the frequency of notable DNA shifts being one in a million, with the enormous bacteria population out there, this tiny percentage is not to be neglected. Hence, at the exorbitant rate at which bacteria recombine, increasing bacterial resistance might become a tremendous issue.

Just paint this image in your mind: bacteria, one of the greatest threats to the existence of all living things in general, gaining total immunity against every single one of our remedies. We would be left speechless in front of this danger, with virtually no way to fight back and defend ourselves. We could try to avoid it, but it wouldn't be like some random epidemic. We would not be able to contain it as it may come in chunks all over the place. Judging on our previous reasoning, this would be, by far, the worst pandemic of all time.

Everyone would get it at some point, and there is currently no prevention or defense method. The only way it would end would be to race through it until we get immune ourselves. That is, the half of us that would remain, or something. It would be the wildest mechanism of natural selection Mother Earth could come up with. And not just for humans, but animals too. Leave the societal collapse for another time, this is enough horror for a scenario!

A disease like this will make Ebola and the Black Death, even Influenza seem tame. Only we don't know when that will come. It could be in the next 50 years, or 1,000 years from now. So, can't medicine figure out literally anything in that time? Perhaps it can. Just like viruses and bacteria have a prospect of modifying their composition, and just like it is within human power to engineer new pathogens, it may also be possible to modify our genetic code to defend against them. This might do wonders for us in the general case of an apocalyptic disease, but it's currently a far cry for our civilization. Because of that, if, by pure chance, total bacterial resistance is reached, or if some kind of potent virus comes into existence in the near future, we'll most probably face imminent disaster.

The walking dead?!

Did you think a zombie apocalypse was omitted from this book? Think again. And due to its sheer distinctiveness from any other type of pandemic doomsday, I felt it right to award it a special place.

A zombie refers to a seemingly revived corpse or a decomposing body which starves to pass on its disease to others... who am I kidding? Everyone knows what a zombie is. Popular culture has famously spun around zombie apocalypses for as long as I can recall, and they reached a degree of popularity way above that of other diseases and even any other apocalypse. It might come as a surprise though, that zombies are not an invention of our developing, globalized world. The idea of inanimate cannibalism has been around for way longer.

The very origin of the word "zombie" comes from the Haitian "zonbi". In Haitian folklore, a zombie is a dead body reanimated through various methods, most commonly magic like voodoo. The trope of zombie cannibalism is thus very ancient, the sole purpose of these soul-lacking bodies being to bring others to the same fate.

Early depictions usually included these elements of the supernatural – dead bodies having a spell cast upon them and reviving to destroy the civilization. Modern media portrayals of the reanimation of the dead don't involve magic anymore, but rather science fictional methods. Corpses rising from their graves have been replaced with more reasonable scenarios, from scientific accidents, to radiation and all the way to the most common one, pathogens acting on one's brain. This is how it relates to our pandemics.

The myth of zombification has deep philosophical roots, some that can be traced back to many spiritual works. The fact that the zombie is insistent on the illusion of the free will, that there is no such thing but only a driving force, silently manipulating us into performing meaningless actions. It is an image of nihilism and idiosyncrasy taken all the way to decomposition. So interesting how the zombie is subliminally described as a harbinger of contemporary nihilism: it has no magic and no meaning like other monsters do, it's simply an inanimate corpse wandering aimlessly in what it looks to it like a desolate world. But as fascinating as this philosophical discussion gets, we shall focus on the scientific side of things for now.

I hate to break the news to you, but a zombie disease is by no means just science fiction, it is real. Of course, there is nothing quite like some apocalyptic movie outbreak that we know of, otherwise everyone would go crazy about it. But something more elusive could be a predecessor of such.

Remember when I said that Ebola was the second deadliest disease one could get? Well, that exception is none other than a real-life zombie virus called Rabies. You may have heard of it, thus it's not the most obscure of diseases, but I personally didn't know how it played out until a few years back. For some context, rabies only claims 60,000 lives per year, mostly in regions of Africa and Asia where it's most common, but the mortality rate once symptoms appear is an astonishing 99.99%!

Basically, the Rabies virus is not a virus like all others. It is firstly remarkable for its extremely wide host range. In the wild, it has been found infecting many mammalian species, while laboratory studies have confirmed that it can also infect cell cultures from birds, reptiles and insects. Almost anything can be a carrier and potentially a transmitter of this terrible virus. However, its resemblance of a zombie virus drives from the way one gets the infection.

Rabies transmission can occur through the saliva of animals and, a bit less commonly but still, through contact with human saliva. It doesn't spread by air or food, but only by direct contact with saliva. This most often involves an act of biting, exactly how a zombie disease is supposed to spread. Additionally, patients who present symptoms are often left behind, isolated from the rest as they will most probably perish in a couple of days. This adds to the reason that some present deviant behavioral manifestations, including agitation, paranoia and, curiously, hydrophobia (the fear of water). Many are tied up to protect others from latent aggression, which can lead to biting and further transmission. The virus effectively acts on the victim's brain and neurologically terminates it.

That's just like a zombie virus. So, how come there is no panic? Where is the dystopian society one expected? There are a few drawbacks to Rabies which didn't let it live up to fictional expectations. The main thing scientists seem to be struggling in every work of fiction entailing a zombie apocalypse is the urge to devise a vaccine. Thankfully, we do have a vaccine against Rabies, which can be administered immediately upon being bitten by some infected animal (dogs, bats, foxes are among the most common).

Then, there is the increased incubation period in antithesis to the reduced symptomatic period. It may take anywhere from a few days to a couple of years to present any symptoms, and you have all that time to vaccinate. Once symptoms appear, no vaccine or medical treatment in the world can save one from death[5]. This takes from 2 to 10 days, and this is the risky time interval in which others may get infected. But because death occurs quite promptly afterwards, there is not that large of a window for other infections. A true zombie virus should be the exact opposite of that – no viable vaccine, a short incubation period and a prolonged symptomatic one. If those conditions are met, no matter how hard you'd try to isolate the virus, it would have time to spread widely. Intuitively, here's where the other factors come in, those being an appropriate place of origin and perilous mutations all throughout (so that vaccines are rendered ineffective and regional quarantine isn't a solution).

Now, viruses are most likely to lead to a zombie apocalypse, thanks to their ability to generate new variants all the time. But a zombie outbreak could also spring from prion diseases. This is an entirely different class of pathogens consisting of fatal proteins which act on one's brain. Most of those occur in animals, probably the most famous being the mad cow disease. But sometimes, those can take different forms in humans as well – the variant Creutzfeldt-Jakob disease being a derivation of the mad cow disease, for instance.

The death sentence is set differently for every prion disease, but during the period following the infection, one develops all sorts of symptoms, from difficulties in thinking and memory loss, to changes in gait and impulsive behavior. Again, it does look like a zombie disease from this description, but there is no evidence of spread via biting, rather by blood, and cases are too sporadic to be considered an outbreak at any point.

So, like other pandemic scenarios, a zombie apocalypse is constrained by the evolution of rabid viruses (Rabies-like), as well as other demographic and societal factors that might come in after the appearance. But it's funny to think about how our world would be like if real zombies were an actual thing; a monster of the masses, carving humanity's way into oblivion. Some more philosophy, to end things on a symbolic note...

Conclusion

In spiritual works, plague is among the most formidable of misfortunes. That makes sense, considering the lack of prediction abilities within our possession whenever a new disease pops out of nowhere. Clearly, it is also one of the most employed in biblical manuscripts as a reference to imminent danger or vengeance. And the idea that plagues have tormented our civilization since its beginnings gives us some insight into the most probably literal uses of the term throughout history.

As of the Bible, I would dare say plagues are one of the central points of disaster forecasts. Often coming accompanied by war and famine (commonly the structure "sword, famine and plague"), we are provided with a means of placing the three at the forefront of biblical prophecies. Although, it is not always just a prophecy, but an actual thing, building up to a lot of bloodshed and misery. However, we are mainly interested in "the plague" that, alongside other potential doom-bringers, would put our world to rest, as it is implied.

In the Book of Ezekiel, the four major terrors were listed as "sword, famine, wild beasts, and plague" (this also comes up in Revelation, right after the introduction of all four horsemen). When comparing to the four horsemen of the Apocalypse as presented in the final book, we notice that only three of the hazards have an implicit harbinger assigned to them, that is Famine (the same thing), War (the sword) and Conquest (the wild beasts) – plagues are seemingly left out. But as we earlier and later saw, they are indeed a crucial factor of the last days of Earth[6]. This might just suggest that plague corresponds to the remaining horseman: Death; the last and most brutal of the harbingers. It may connect even better when thinking about how "Hades was following close behind him", possibly signifying the way in which a subsequent event will be as brutal as the plague itself.

Of course, terms like "pandemic" and "virus" have no reference in neither the Old, nor the New Testament. On the other hand, intriguing sequences could be interpreted as early signs of the end times.

"But after they had moved it, the LORD's hand was against that city, throwing it into a great panic. He afflicted the people of the city, both young and old, with an outbreak of tumors."

1 Samuel 5:9

"They have power to shut up the heavens so that it will not rain during the time they are prophesying; and they have power to turn the waters into blood and to strike the earth with every kind of plague as often as they want."

Revelation 11:6

Depictions of diseases also get quite explicit at times, with some of them known to us and explained in close detail such as cholera. But there is no evident description of the final disease, the one which may kill up to a forth of the humankind. Perhaps the unpredictable nature of pandemic is behind all this. Maybe there will be not one disease to dismantle our civilization, but multiple outbreaks around the same time. Something of these proportions, whilst less likely to happen, would pose a serious issue to epidemiologists looking for cures and remedies, which might just be enough to reach the amount of devastation showcased in the Bible.

Realistically speaking, an apocalyptic pandemic may not be as odds-on as one would believe after finishing this chapter. But reported to the rest of the analyzed scenarios, this one has a gargantuan probability of happening. The fact that no pandemic has reached such a potential yet may be a bit encouraging, but then again, our world is completely different now. In our efforts to develop as a civilization, we've also constructed the ideal environment for the spread of plagues. This reason is, on its own, terrifying.

We are constantly playing the pathogen lottery. The chances are slim, but not nearly as slim as in normal lottery. It's more of a Russian roulette really. And who knows when the bullet will come out...

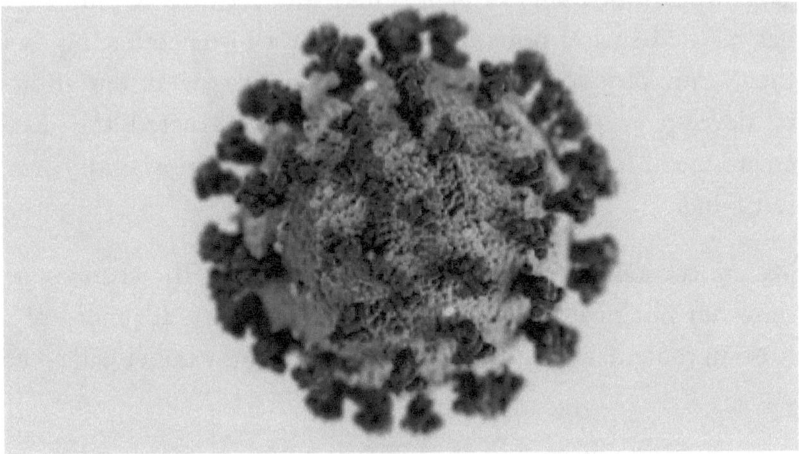

The COVID-19 pandemic has forever shifted our modern perception of diseases: we took one for the rest of humanity, hoping that we will be more prepared in the future

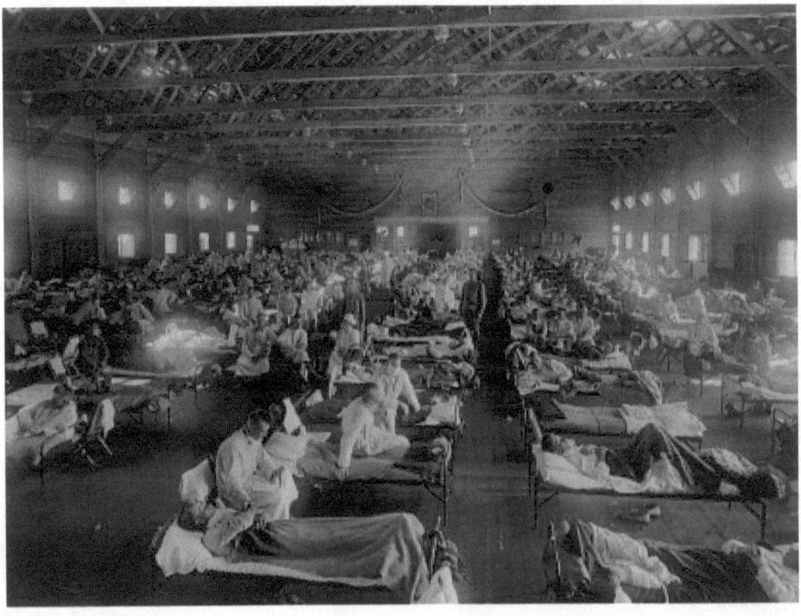

TRAIAN TUS

A hospital ward from Kansas, filled to the brim with patients suffering from Spanish Flu

1. Some diseases do not have a 100% efficient remedy or treatment, COVID-19 being one of them, while some have no known cure at all.
2. "Ground zero" refers to the starting point of the infection. Just like it represents the source of an earthquake, let's say, in this case it represents the source of the disease. "Patient zero" refers to the first person who contacted the illness.
3. The virulence of a disease is the biological term for its severity. If a virus only results in coughing and sneezing, then its virulence is not that high, but if its symptoms include more serious issues like pneumonia or skin lesions for instance, then the virulence would increase.
4. Referring to Influenza, an antigenic shift describes the emergence of a new strain of the virus, resulted from the recombination and reassortment of two different Influenza viruses.
5. There is only one patient known to have overcome Rabies without seeking any medical attention or being previously vaccinated against it. All the rest have died.
6. In that context, "Earth" actually means "humanity", as humans are concerned, not necessarily our planet.

IV.3: Nuclear war

"The measure of a man is what he does with power"

Plato

After countless doomsday scenarios analyzed, one more terrifying than another, one likelier than another, we've finally made our way to the very top of the apocalypse ladder – the peak of the mountain of horrors. You probably saw it coming by now (even if you somehow entirely missed the contents section): that the most notorious possibility for bringing the end of our civilization is none other than a global self-destruction event.

Time and time again, yet still ringing loud as ever, perhaps the one feat we are to be most frightened of is us. The very thing we are meant to protect might prove to be our greatest enemy. An intrinsic threat, dwarfing all of the others, especially due to its outstanding volatility and unpredictability. It's funny how, after all these scientific apocalypses that we've covered over the last few hundred pages, we've come to grips with the compound nature surrounding us, and still have no idea about the basic nature within us...

But basic as it may look at first sight, minuscule when compared with what the rest of the universe has in store, the human nature is actually of other-worldly complexity. It doesn't always make sense, unlike the physical nature. Rather, it has a slight subjective touch, which, in most cases, might reach crazy proportions. This is by most accounts harmful on all planes, be it social, emotional and even physical, but thousands of years of human evolution have showed us that it's something we have to live with, something we seemingly can't get over, and which sparked who knows how many disputes over time.

Still, we couldn't blame any past apocalypse on ourselves. Whenever time came for a mass extinction event to ensue, it was without exception caused by natural phenomena. Once it was an asteroid, once it was a super-volcano, once it was climatic changes throwing Earth into an ice age, but it was never just humans that led to such a wide-scale disaster to imperil our existence. And that's not because of our compassionate side, as that isn't quite dominant on certain occasions. It is mostly because of the lack of potential at any given prior moment.

One may think it used to be easier to erase some tens of thousands off the face of the planet than it is to do so with eight billion and going. But really, it might be somewhat of the same difficulty. Thousands of years back, those tens of thousands of people were dispersed all over the place, living in isolated communities of tens to hundreds with little to no interaction with the others. It took a decent while for this to change and for society to be formed. It was only with the late development of agriculture that something of a city could come into being, when larger groups of people coalesced into bigger and more advanced establishments. That meant a better way of living, but also a higher risk of destruction in terms of human lives. If a volcano were to erupt or if an epidemic suddenly struck in those early times, almost certainly every citizen would suffer from it. But that didn't mean it was easy for humans to self-destroy even one such community. The technology was rudimentary, and there were other things to worry about. Much sooner would an unknown illness kill you and your family than a random lunatic impaling everyone with his wooden spear.

Things have gotten darker and darker since. One might argue that it only got brighter due to our rapid development in all areas of technology and society, and that would be correct. But particularly apocalypse-wise, we are much more susceptible to a unanimous devastating event. We now have the power to take a lives just as easy as any other mass-extinction event. We are able to fiddle with nature in such a way, that humans don't stand a chance against our own inventions. Engineering viruses, changing the climate and the environment beyond recognition and even stimulating all sorts of natural phenomena could well harm the entirety of the human race. But in this chapter, we are going to focus on the most reckless of inventions, the ones capable of destroying the world at the push of a button, and where human nature plays the most vital role: nuclear devices.

The (new) art of war

Much is known of the homonymous work attributed to the Chinese general Sun Tzu. For centuries have his advices and techniques been used in warfare. In spite of me not being able to consider something as horrifying as war an art, I have to acknowledge that it takes exceeding skills to perfect military strategies and procedures. And while this is pretty obvious for any type of confrontation provided one's need to get the upper hand over his adversary, something has drastically changed since the time of Sun Tzu.

The predominant idea behind modern armed conflicts has nothing to do with the general's teachings – it doesn't focus on the "art" of war. Instead, it is geared towards military superiority. And I'm not talking about man power (how much military personnel each army has at its disposal), that is a thing of the past. It is about the military technology available to the belligerents. Better equipment, improved offensive, enhanced machinery and you can see you opponent's chances of coming up victorious essentially plummet. He could know by heart all of the best maneuvers and tactics ever used in combat but in vain, he would be crushed like an ant by your lofty technology. Just take the battle of Rorke's Drift for example: 150 British troops defending an outpost against 4000 Zulu warriors. For all their numerical superiority, the wooden shields and spears of the Zulu stood no chance in the face of the Brits' rifles. Now imagine that, taken to the level of robots, drones, missiles, supersonic jets and what not. Oh yes, and how could we forget about nuclear bombs?!

Ever since the beginning of World War II, the arms race between the world's superpowers was relentless. Everyone was competing to achieve higher military potential, but it was in 1939 that the quirky, yet fearsome atomic bomb was conceived. In light of previous discoveries in the field of nuclear physics and the growing fear of the German army, Albert Einstein suggested president Roosevelt that Americans begin the development of nuclear fuel, envisioning a weapon without precedent. Soon afterwards, he very much regretted his action as he fully understood the devastation such a device would inflict on human population. He called it his "one great mistake" and later noted: "Humans invented the atomic bomb, but no mouse in the world would construct a mousetrap.". Though it was too late. He warned that "we thus drift toward unparalleled catastrophe.", which certified in August 1945, when two atomic bombs were dropped on Hiroshima and Nagasaki, in Japan. 225,000 is a reasonable estimate of the number of lives lost in the two bombings. Small having in view the total of 40 million casualties of WWII, but phenomenal for a single device. And so, the arms race has commenced.

At first, there were these two and another nuclear device also set to detonate in Japan in case they didn't surrender to the Americans after the first two. But in the years that followed, the USSR gained ground in the arms race by devising their own nuclear weaponry. The Cold War was an important milestone as tension has been building up between the great nuclear powers. Both reached an apogee in terms of warheads and they came worryingly close to a nuclear exchange towards the end of the war[1].

It kept on going for a while, until a number of scientists complained about the danger they pose to our race and demanded they are all disarmed and that further production is prohibited. A number of prohibition and non-proliferation treaties have been signed with respect to these complaints, involving almost all of the world's nations. But as usual, we aren't completely safe in this regard either.

I mean yes, you can sleep much safer at night nowadays than you would at the height of the cold war. You aren't taught in school how to survive a nuclear attack and you are not instructed on how to build a bunker on TV, so you know it's not that bad. The number of active nuclear warheads has significantly dropped in the past 50 years. But we live in a perilous era: a bit over 15,000 are left active to this day.

Spread across 9 countries of the world (although more than 90% of them belong to Russia and the USA), any single nuclear detonation would spark endless quarrels among the nations. If, in addition, one is detonated on foreign territory, a like response is almost inevitable – the game of nuclear football is on.

You may be thinking five, ten, let's say 20 bombs are detonated in this kind of exchange. But you would be amazingly off. Once the first bomb goes off, a full-scale nuclear war is born. Everyone would be dragged into this kind of conflict, and no one will come victorious. There would be no winners in this type of war. Every person, every country, the human civilization itself and our nourishing planet as a whole will severely lose in such a scenario. An apocalypse is inherent to a nuclear war, and this is the primary reason why we must avoid nuclear weapons at all costs.

But for all sakes and purposes, before diving deeper into the hell that would be unleashed by a nuclear exchange, let's assume that a single device is triggered. What would follow and what are your chances of making it out alive?

Kaboom!

Say you are somewhere in New York, not necessarily living there, just on a business trip, to make things more challenging. All of a sudden, your quiet walk is disturbed by an uncanny noise. You soon make out what's going on: this is not an exercise, it is the real Emergency Alert System warning everyone that something bad is about to happen. You frantically look left and right to find shelter, but time is very short before it wreaks havoc. A nuclear bomb has been dropped and detonated over New York! From now on, it's more about luck than about your enviable physique and survival skills. A few cases are possible in this situation.

First off, if the nuke is detonated during the day, a sudden burst of light would engulf the city for a couple of seconds, immediately upon detonation. After that, the air would look yellowish-orange as the giant characteristic mushroom cloud[2] would begin to rise. If you happened to look in the direction of the blast, temporary or permanent blindness would take over you (depending on your distance from it). Actually, you would be bewildered and your vision hazy wherever you looked at the time of detonation, but the effect wouldn't be as bad. If, however, the bomb went off at night, the difference in brightness would be even more disrupting to your eyes, so it would be better if you were already inside.

The height above ground level is also important for a detonation. Usually, and by that, I mean the two WWII bombs and most of the tests that followed, nukes are detonated above the surface in what's called an "airburst". The advantage of such a detonation is making the most of the effective blast. That is, more explosive damage thanks to buildings not standing in the way of the blasts. Additionally, most of the radiation[3] would be propelled into outer space and not remain in the atmosphere to cause extensive radiation poisoning as it comes back down. In the eventuality of a surface burst, the fireball wouldn't be as big due to buildings and land formations obstructing the area, but the radiation would have nowhere to escape and it will thus persist in the atmosphere. The wind plays a vital or, better said, morbid role here, for the radioactive particles would follow its direction and spread all over the place as they fall down.

Now back to the explosion. We'll take the worst-case scenario, when it's rush hour in New York and the nuke goes off high enough to yield maximum explosive damage, yet low enough for the majority of radiation to be retained in the atmosphere. Anything from windspeed and wind direction to other atmospheric and technical parameters can alter the result, but we'll consider average conditions and a 100 Megaton detonation (the most powerful nuclear bomb designed). It is worth noting that in a global nuclear exchange, it wouldn't be just 100 Megaton bombs, but probably much more smaller ones of a few Megatons each (B-83s, the largest in the current US arsenal have around 1.2 Megaton of explosive power, but more powerful nukes are also available). Yet for this single detonation over New York, we'll play around with the most fearsome creation of humanity.

As it goes off, a huge fireball is generated, practically vaporizing anything situated within an 8-kilometer radius from the blast. A shelter wouldn't help if you happen to find yourself at such a distance. The direct radiative area is similar in size, with strong radiation reaching you almost instantly (as it travels at the speed of light). But don't worry, your receptors would be fried faster than you could perceive anything so it would be a swift death. This radiation is very energetic and can penetrate through multiple layers of protection and affect the entire living tissue inside your organism (gamma radiation especially). A few feet of concrete between you and the outer environment would be compulsory for survival. Even after the explosion, this central zone would be most heavily affected by subsequent fallout. 10 kilometers away from ground zero, heavily built concrete buildings are severely damaged or demolished, while fatalities approach 100% inside this "heavy blast damage radius". Moderate blast damage goes as far as 20-25 km, with most residential buildings collapsing, and up to 55 km away you can expect windows shattering and light damage to structures. You may consider yourself lucky if you were situated farther than that (but you wouldn't really be in New York anymore). Still, you might want to go as far out as 70 km from the blast, otherwise you might be exposed to thermal radiation, leading to third degree burns, extending throughout the layers of skin. Yet again, you wouldn't fell the pain as the pain nerves would be destroyed in the process, but you can imagine how bad it would be anyway.

You are safe from the blast, you are safe from the initial radiation wave, you only have to get over the fallout. Hopefully you are not short of breath yet, because the struggle has just begun. Up until this point, it was whether you were lucky or not to protect yourself from the blast. This is where your knowledge would come in handy.

If you weren't that far away from the explosion, you will want to stay in your shelter for two or three days until the fallout settles. But first and foremost, you want to get rid of your clothes and wash your body and hair to get rid of the radioactive particles that have clung onto them. Meanwhile, you should keep an ear out for any rescue coming, if your radio station does, in fact, work[4]. But don't venture outside just yet. Not without protection, as the air is still moderately contaminated. Your best bet is to just be patient and wait for the military to come and help you out.

If you were situated at a considerable distance from the bomb, even in another city (e.g. Boston, 300 km away from New York), you would want to study the direction in which the wind blows. You will probably figure it out from the news that will web throughout the American television. This way, you can evaluate how the deadly hail will be carried in the upcoming period. This is no place to underestimate the severity of a nuclear bomb. The fallout is generally much more dangerous than the fireball and tackles a much greater area. Just portray an area of 50,000 square kilometers affected by radiation rates of 1,000 rads per hour, and an area ten times as big for high, moderate and light fallout. Doses of 100 to 200 rads delivered in less than a day can cause acute radiation syndrome, but are usually not fatal. On the other hand, 200 to 1000 rads accumulated over a few hours cause serious illness and poor chances of survival, while 1,000 is no match for the human body and are almost invariably fatal. You want to avoid the affected areas as soon as possible, travelling perpendicular to the wind direction to get out of the hotspots.

Fingers crossed you don't die of cancer in the next years, you managed to make in through a nuclear explosion safe and sound. You had a better fate than at least 10 million others. It is far-fetched, but not impossible. As a matter of fact, there is one person about whom we know to have survived both the Hiroshima and Nagasaki bombings while witnessing them on the scene. So why wouldn't you be able too?

The aforementioned scenario is a hypothetical one, like many in this book. From a physical standpoint, it is a fair approximation. But in reality, such a bomb wouldn't be detonated on its own. A prompt reaction from the world's nations is more than certain.

Embrace the chaos

If there is any one true definition of "mayhem", this is it!

The very second an ICBM (intercontinental ballistic missile) is detected – because nowadays nuclear warheads are launched from a distance, they are no longer carried by puny aircrafts above their intended target – retaliation is on the way. This would further increase the tensions and all of the allied countries will get involved in the confrontation. Nuclear powers or not, everyone will play their part in what could be the last stand of humanity...

While the outcomes of such a conflict are tough to predict, some of the scheming ideas of the past that have recently seen the light account for thousands of bombings worldwide, even in places which are seemingly not included on the war maps. The fits of rage of any nuclear country could cause all nations, even the non-nuclear ones, to come to blows.

The story tree branches wildly. Half a century ago, all these proposed bombings would have accounted for one to at most three billion deaths worldwide. But as we know, our war capabilities have gotten way nastier since. No one knows how many people would die if something of this magnitude erupted today. Nonetheless, we could guess what would cause the succeeding chaos.

The single-bomb-scenario is not the end of the world. It is, clearly, something so macabre that one shouldn't even think about, but society will rebound. Every building within 20 km or so would be levelled off, so what? With the current technology, we could rebuild everything from ground up in a few decades, as the fallout would eventually calm down. The two Japanese cities look much better today than many that remained intact, as if nothing ever hit them.

In the several thousand-bombs-alternative (the more realistic one), we wouldn't have such an easy time recovering from the devastation. It wouldn't be barely one city hit by a nuke, plenty more will follow in just a couple of days. Nukes will sometimes overlap, making sure that important facilities and urban formations are fully covered. Those unfortunate enough to be situated in the targeted areas or even somewhat close to them (somewhat close may mean 1,000 kilometers or more) will stand little to no chance of making it through if they don't act precisely right away. Even so, tens of millions to even hundreds of millions will die in the most populated countries from the initial blasts only. The nuclear fallout would be drastically increased in contrast to the one theoretical warhead dropped on New York, encompassing vast areas of land, and making it difficult for help to arrive. Also, you might not even rely on radio or any other kind of electronic devices, because strong electromagnetic pulses (EMPs) are part and package of every atomic detonation – remember how they would fry anything in their path from previous chapters (maybe only those sturdy meters of concrete would shield your appliances). In this scenario, you may sooner die of hunger or thirst than for the military to rescue you. The world would be collapsing before our eyes and everyone would be on their own.

The fallout will need some time to quench, taking the lives of some other billions by then. Extensive wildfires will, over and above, roam the planet as they are ignited by the thermal waves of the explosions. They would have no time to burn out before they spread and virtually anything in the way that can catch fire will catch fire. Quite a hellish image should form in your head – all those explosions and fire and toxicity. But possibly unexpectedly, something diametrically opposite to that is about to be set off.

If you can recall, we've had a discussion about the aftermath of volcanic eruptions, in particular, the cases of super-volcanoes. Upon eruption, thousands of cubic kilometers of volcanic material are spewed into the atmosphere and are trapped there for years to come. They spread across the globe and obstruct sunlight, leading to serious drops in temperature, changes in climate and weather patterns and, consequently, to all-round famine and loss of lives. Now think: what does this resemble? Hmm... oh yes, nuclear explosions.

The mechanism of a nuclear detonation is by no means like a super-volcanic eruption. Conversely, the nuclear clouds of exhaust material are very much akin to those of volcanic ash and dust, in the sense that they behave similarly. Both are toxic in their own way, that's one aspect. But the underlying feature is that during their prolonged stay in the planet's atmosphere, they block sunlight from reaching the surface. One thing leads to another and we are stuck with a mini ice age for a few years to decades.

This "mini" word might not accurately describe the gravity of this scenario. For instance, the Lake Toba eruption 75,000 years ago left Earth in a decade-long winter, with temperatures 5 degrees less than the norm. Research suggests that the respective period was a crucial point for the human race, as only 3,000 to 10,000 individuals we still alive after the icy season ended – a genetic bottleneck[5] in human evolution.

Such a reduction is improbable in our 8 billion individuals world, but a similar event may usher in a much greater death toll. In their work "Climate and smoke: an appraisal of nuclear winter", astrophysicists Carl Sagan and James Pollack studied the physical effects a nuclear war would have on our planet. No surprise, they reached to the conclusion that it would heartily disrupt the global climate. By vaporizing debris and filling the atmosphere with ash, soot and aerosols, a nuclear war involving thousands of warheads might easily cause a cascade cooling effect, which would plummet global temperatures by as much as 22 degrees! That's... a lot.

Although their original estimates might have been a bit overzealous, it's key to understand that our environment is incredibly fragile – even a ten-degree drop would have dramatic repercussions. Apart from obstructing solar energy, something along those lines would send weather and ocean current patterns that help keep the world mild today into disarray. Nothing would be the same in this man-made ice age. In the lack of sunlight, you would probably be delighted to see a few shafts of it through clouds full of choking dust - but beware that sunlight could be lethal. The ozone layer would have been radically damaged by then, so basking in the sun for too long will result in severe sunburns and cancers (if the explosions themselves didn't give you enough trouble with that).

And the thing is, I am probably still short of fully emphasizing the situation. When we think about a nuclear war we imagine the big cities being targeted and destroyed, and that in this post-apocalyptic society we can make it through provided that we learn how to grow crops in extreme conditions and stay isolated, safe from all the chaotic dangers. But it isn't just like that. It would be too simple. You would have no time to react, and no way to fully prepare. An interesting simulation has been created, focusing on a potential war between India and Pakistan (isolated, that is, with no other countries taking either side). It would take only 2 weeks for the smoke to engulf the planet – a framer maybe in Europe or in the US, many thousands of kilometers away from the incident, would look up at the smoky sky above him and down at his crops that have died in his field from lack of light and cold temperatures. An all-out war would be astronomically worse than that.

Owen Toon, professor of atmospheric and oceanic sciences who has studied nuclear war for 40 years, estimated that "90% of the world population would perish, and civilization would be destroyed.". That's how bad it could potentially get. Frightening still is the fact that this isn't just another doomsday scenario, hidden among the others. The world is "prepared" for it. Not for its consequences, but for setting it off. Truth be told, "if you live in a city that has a military base, there is a missile that is aimed at you right now; if you live in a city that has an important industry, a major university, a large airport, an oil refinery or oil storage facilities, there's a hydrogen bomb that's aimed at you right now.", said the professor. Our first premise was to destroy the world, whereas the second premise was to prepare for it. The whole idea of being safe in a nuclear war is purely delusional. It wouldn't help you to duck and cover. Not even a highly secured bunker would do much of a trick. We are all in the same boat, and it will be sinking fast...

A war on all fronts

But if only it were that easy... You'll say: "Yes, sure, but there must be some kind of defenses in this scenario. Some kind of failsafe like it was the case for any other apocalypse.". Well, I'm sorry to disappoint you, but there isn't. There is no way to foresee what's going to happen in a nuclear conflict, let alone when one is doomed to start, and there is no way everyone can be protected from the raining nukes.

The fact that the number of nuclear warheads has dropped about three times since the height of the Cold War is somewhat soothing. However, as the technology advanced, so did the potential of those devices. Intercontinental Ballistic Missiles are, as I've said, the go-to when a nuclear attack is intended – just select your target and it's already on the way. With novel supersonic advances, any detection system will have an even harder job to identify incoming missiles.

But those have gotten better too. Hundreds of cutting-edge detectors span the globe, achieving 100% coverage of latent ICBMs once they reach a certain altitude. As soon as that happens, national defense systems of allied and especially targeted countries are alerted, for a lightning fast neutralization of the nuclear threat. An interceptor missile is set to tail the ICBM and destroy it without triggering a similar explosion as the deliberate one.

As reliable as those might be, you never know when a faster-than-sound projectile passes the barrier and manages to enter the vicinity of the target location prior to detonation. Reality has it that ICBMs would therefore not come on their own. The attacker would want to make sure that the target is hit, therefore multiple warheads would be sent our way, with a view of at least a couple of them making it through. That would be enough for hell to break loose. Above that, even if all of the nukes are somehow neutralized in time and none manages to get their job done, the sole intention of bombing a specific territory will be accompanied by a like response. The origin of the missile would quickly be detected, and retaliation from the "averted victims" is quite certain anyways.

Intuitively, a nation's main issue in such a moment of despair would be to get rid of the threat in the shortest time possible. But retaliation may come even before the incoming missiles are put down. Letting conventional detection systems aside, there is another efficient way of making out a shot ICBM: satellites.

With those celestial watchers surveilling every inch of our planet 24/7, you get an idea of how fast a nuclear missile could get detected after launch. This way, a considerable heads-up could be received by the targets in order to prepare and neutralize the danger in due time. But, believe it or not, even satellites may prove erroneous at times and, worse still, they could be neutralized much easier than an ICBM.

You've got it by now that a nuclear war would not be any kind of war. It would be much more a war of patience, manipulation and gambles among the world's leaders than of tactic and strategy like they used to be. The deal is significantly more complicated before the first weapon is shot. After that, everything goes crazy. And it is in that waiting period when what could be called "the strategic action" happens.

In this amalgam of earth-shattering events, the priority would be to make your adversary as weak as possible before the real danger kicks in. For this reason, we would assume that a nuclear war isn't summarized to nukes, drones and satellites, but also other methods of warfare. Among those are the previously addressed notorious scenarios involving cyber and bio warfare. It is superfluous of me to explain what would happen if all these horsemen joined together in an "effort" to destroy our planet. Let's just say they have astounding chances of success.

Remember how the satellites used to be an important tool in the detection of ICBMs or other military threats? Well, those wouldn't be of much use if bombarded themselves or if simply shut down by groups of hackers. Whether they would be assigned by the government or simply groups of fanatics taking advantage of the collapsing world to further mess with it doesn't make that big of a difference. But you can imagine internet, television and essentially any other way of communication going down during a nuclear war. Anything else from the fall of infrastructural systems to the inefficiency of currencies might also be traced back to this side of the war. At the same time, biological weapons could be employed by any nation or terrorist group. After all, there are no more rules in this kind of war...

To sum it up, this war would be fought on all thinkable fronts in order to ensure maximum advantage. Really, it would only ensure maximum destruction. As with all apocalypses, we're all in the same team, though in this case, countries seem to be more antagonistic than ever. The only true opponent is the very existence of life on Earth.

After the invention of the atomic bomb, Albert Einstein was at a dinner party with a group of friends and, all of a sudden, he was asked what new weapons might be employed in World War III. Appalled at the implications, he shook his head and, after giving it a thought, he said: "I don't know what weapons might be used in World War III. But there isn't any doubt what weapons will be used in World War IV.". "And what are those?" – "Stone spears.", or in other versions, he was quoted as saying "sticks and stones". Sticks and stones everyone... let that sink in...

Nuclear terrorism

What if, by chance, we make do with our current tensions and leave the past behind? No historical, nor political motives to start a war of these proportions in the first place. Surmounting the hostile human nature, as doubtful as that may sound, and understanding the nature of conflicts in order to avoid this apocalyptic prophecy. So, say we are living 50 years from now, when all of today's mundane issues have been overcome and no altercations between nations even have the chance of escalating to nuclear. However, nuclear weapons are still present in countries' arsenals just in case, or simply for other purposes like deviating asteroids or using the energy as fuel.

Yet out of nowhere, a nuclear device goes off. What could it be? Who had the audacity to strike with such force without any prior warning? No one knows. At this point, the animalistic spirit within the rational humans of the future shows up again. After decades of peace, the tensions between nations reach unseen levels in just a matter of hours from the explosion. Some are trying to calmly analyze the threat, but others start pointing fingers right away. Each of the nuclear powers blames some other. The heat rapidly goes up until the last straw: some country decides to launch a nuclear attack, in an attempt to forcefully reply to the first strike. Only later does some terrorist group take responsibility for the nuclear attack, but it's too late... a nuclear war has already begun.

Of course, it isn't implied that a nuclear war would start from a terrorist nuclear attack, the situation described above is just one of the many ways things could unfold. What's clear, is that a single terrorist attack of this sort would have terrible consequences. Now think about more of those.

We would have to think differently in the case of a terrorist nuclear attempt. The principles the nations would abide by even in the case of a nuclear war are virtually inexistent here. The one and only reason for a nuclear attack like this would be to cause as much damage. And, in the XXI century, this could be a real threat, possibly bigger than an "organized" nuclear war.

Not only is it possible, but it is really probable. It's actually shockingly feasible to develop a nuclear device even without the top facilities of certain governments. While the development of each country's nukes is classified, there is a lot of know-how in the area of manufacturing improvised nuclear devices. Everything is out there: detailed information on how to assemble a nuclear weapon from parts, how to build a nuclear bomb, how to create a terraform where you could actually manufacture and create all of the components – the only difficulty is getting one's hands on the nuclear fuel.

If this obstacle is surpassed, it isn't long until a nuke is assembled from scratch. The way the containing and triggering mechanisms are devised might vary, but the core itself (the part which undergoes nuclear fission) needn't be that massive. Some tens of kilograms of uranium, neptunium, plutonium or other elements would be more than enough for guaranteed devastation. Adding the weight of the rest of the contraption and everything could be fitted in a truck or a larger suitcase. Obviously, it wouldn't compare with the most powerful of ICBMs, but it would be bad enough for chaos to ensue. One day, a casual-looking truck could be parked somewhere in downtown New York, the next, half of the city would be wiped out.

And there isn't only the problem of extreme unpredictability. What's worse is the world's inability to fight back. Terrorist organizations are stateless. Their members are spread across the globe and vile groups could be planning their taking over at any time, without anyone's awareness. So, if one happens to attack, how would we strike back? Would we simply bomb another nation because we guessed that they are hiding there? I don't think so. It would be pointless. A year down the line, something similar could occur in another location and be the credit of the same group we thought we eliminated a while back. Also, what would even mean to retaliate in such a scenario? Those are people willing to give up their lives to inflict damage with nothing to win out of it. It becomes apparent that this whole notion of mutually assured destruction would not work in these considerations.

In this perspective, terrorism is unstoppable, and its nuclear option is not a far cry. Especially in this case, we are all fighting together for the good of humanity, and not the needs of whatever other entity.

Conclusion

A self-destruction event has long been theorized. Be it via nuclear war, the invention of some sort of doomsday device (creating a micro black hole was among the most upvoted possibilities) or another feat that would have a strong, direct impact on the whole of our race, an event like this is widely considered the most probable way humankind will die out.

We saw life bounce back from the poorest of conditions. No asteroid, super-volcano or climatic catastrophe was able to terminate it. Whatever the consequences, we were able to survive the worst. And not just that, but we kept on evolving until we reached the present form. We are at our best in terms of development, yet at our worst in terms of threats. Sure, asteroids might not be a problem anymore and we pay much more attention to ecological disasters, but all the rest... there's still a lot we'll have to deal with. It isn't easy being us, but we were always victorious.

That's a bit strange, right? Looking back at all of our fallacies and misbehavior, it is difficult to imagine a brighter future. Maybe we just got lucky? Though it's even harder to believe how lucky we must've been. I think the real reason is because the worst is yet to come. We haven't yet faced that great danger that will decide whether we are worthy of living on or not. We thus return to the idea of a Great Filter among each civilization's evolutionary journey.

Although the link might not be evident, a quick reminder that this last scenario is intertwined with almost all the others. What's up then? You see, almost every scientist that has ever reflected on this issue reached the same conclusion. There can only be one plausible Great Filter to have acted on most of the civilizations out there, civilizations that had nothing to do with each other, though seemingly reached the same fate (hypothetically). Everything points towards a self-destruction event. That there was, or will be, at least one point in a civilizations development when something will go horribly wrong and would lead to the self-destruction of the civilization. It comes down to their ability to avoid it or, at least, barely make it through, though the former seems more unlikely.

By the looks of things, a nuclear war might be humanity's Great Filter. It is the first time in history when we are fully capable of destroying ourselves in a direct, abrupt way. We wouldn't know until, God forbid, it happens, but the indications are many and incredibly terrifying. If all of the war metaphors and the blood and fire and the worst horseman of the four isn't clarifying enough, then the whole rest of the biblical apocalypse should do, as everything would come down crumbling in the advent of a nuclear war. One particular image is most often associated with this scenario, although quite hard to decipher. But let's be serious, nothing could be easy in the most probable of apocalypses...

"The locusts looked like horses prepared for battle. On their heads they wore something like crowns of gold, and their faces resembled human faces. Their hair was like women's hair, and their teeth were like lions' teeth. They had breastplates like breastplates of iron, and the sound of their wings was like the thundering of many horses and chariots rushing into battle. They had tails with stingers, like scorpions, and in their tails they had power to torment people for five months."

Revelation 9:7-10

The mushroom cloud released by a nuclear bomb can reach tens of kilometers in height

1. By many standards, the riskiest moment in terms of nuclear tensions was defined by the Cuban Missile Crisis in the autumn of 1962.
2. Just like in the image above, a mushroom-looking cloud is released upon a nuclear detonation and its dimensions can be quite remarkable.
3. A few types of radiation result from a nuclear explosion. Alpha radiation interacts the most powerfully with the substance, being absorbed after just a few millimeters. Beta radiation follows, they too are highly energetic particles which pass through thicker layers of the substance, but are still absorbed relatively quickly. Gamma radiation, the most dangerous of them all, consists of high-energy photons that can pass through thick layers of substance, therefore a strong shielding is needed in order to stop this type of radiation from reaching you. In case of an airburst, a large part of this radiation is expelled into outer space, posing no further danger.
4. Due to the electromagnetic burst inherent to a nuclear explosion, circuits might get fried and your radio might not work properly.
5. A genetic bottleneck occurs when a population is greatly reduced in size. The bottleneck limits the genetic diversity of the species because only a small part of the original population survives (and it looks like a bottleneck on population graphs).

EPILOGUE

"Destiny is not a matter of chance – it is a matter of choice. It is not a thing to be waited for – it is a thing to be achieved."

William Jennings Bryan

After all these years since the incident that sparked my interest in the surprising identity of the Apocalypse, I can finally claim that I am at peace. I am now calmly sitting at the desk, with this controversial book before me, contemplating on the good memories that stuck around, and being curiously enthusiastic about what's coming next. And you should be too!

When I began writing this book, a much more sorrowful state had taken over me. I wanted to express and explain not the need for an apocalypse, but the need for closure of the human cycle. Thousands and thousands of years of evolution, with their ups and downs – a period of great success for life itself but, at the same time, a time of unprecedented grief and horrors. It only seemed natural for something of this breadth to eventually come to an end. It was but a question of how and when. Which of the apocalypses would come first and how much time do we have left until the ultimate destiny of our civilization is reached.

As I've progressed through this work, I couldn't help but realize something was changing. Every single day, a bit of me felt different about the motive behind this book. And I believe that, as the reader progresses too, he can feel this distinguishable shift in the text. I am no longer seeing the apocalypse as the end of everything. Rather, I view it as a process of metamorphosis – from caterpillar to butterfly. An unthinkable transformation that would launch us into an entirely different plane of living. Perhaps we are simply not ready for that yet.

I don't mean to sound too radical, but I figured it interesting if I punctuated this connection in essence between the scientific and biblical apocalypses. As far as the Scripture goes, the Apocalypse is not the end of days, it isn't the destruction of humankind. In actuality, it signifies a change in the general perception of things and the revelation of what's beyond our limits. That's why the last book is called "The Book of Revelation", because the hidden reality is finally revealed to humans. Something much greater is awaiting. In the book, it isn't God's purpose to destroy humankind, Its own creation. It just helped them visualize the whole picture.

I would say that this here is a more appropriate approach. Memento vivere (remember to live) instead of memento mori (remember that you will die): that's the great antithesis behind it all. Yes, the apocalypse may come someday, maybe 100 years from now, maybe a thousand, maybe even tomorrow, but that's not a reason to be afraid. You could go through your life wondering "what if the apocalypse comes and it was all for nothing", at the price of actually living the beautiful life that you can create for yourself and others. Don't make the same mistake I did – always worrying about the future and never getting to enjoy the present. You might never be happy that way. Learn to enjoy the little things and make the most of everything. That's not to say that planning ahead isn't essential. Only remember that thinking is more important than calculating, no matter what comes up.

The apocalypse should not be taken for granted. It isn't definitive for our human race. It isn't set in a particular moment of time in our existence. Our fate is eternally adapting with respect to our choices, and if it has to come, then it will eventually reach us.

Meanwhile, we should avoid being drawn back to the legend of Sisyphus and look into the meaning of things. Maybe our boulder will reach new heights that way...

In doing that, we are to consider the aspects presented over the course of this book. We should constantly aim at surpassing our boundaries, but we must be extremely cautious not to bring doom upon us faster than the real apocalypse is meant to come. But then again, it might not come at all, so who are we to decide it does. It isn't rightful to assume anything, by any stretch. All we can do right now is be attentive.

And from what we've seen, even though chances don't seem to be on our side on many occasions, it is well within our ability to change that. We can and we have to play our part precisely on this grandiose stage and everything will be fine in the end. And if it isn't, then it's not the end yet.

Our world is, progressively speaking, better than it ever was. In spite of the plethora of threats to our well-being, the lifestyle of human beings has reached upmost levels of security and prosperity. We are doing so great, that we can't even picture a lesser world. And perhaps this is the reason why we are so scared all the time. The safer we are, the more afraid we get of any potential disturbance of our current, nearly perfect condition.

On these grounds, my opinion is that we are living the most crucial time ever. One could argue that any "present" moment could be deemed the most important, but overall, now is the time for a crossroad in humanity's journey. It is now that we have to take one and only one path that will define our future: do we want to put in the effort to ascend our civilization where we could only dream of, or are we satisfied with our ominously flawed world and go spiraling down to an inevitable doom?

We thus return to my previous statement, that we should be enthusiastic about the future. Whatever happens, we shall get over it. Stephen Hawking put it in a milder, hopeful tone: "Sooner or later disasters such as an asteroid collision or a nuclear war could wipe us all out. But once we spread out into space and establish independent colonies, our future should be safe.". In a nutshell, it is now that we decide which path we take. And it could also be that it is now or never if the apocalypse comes or never does. Because as we develop further and attain multiplanetary or even galactic potential, it becomes exponentially more difficult for an apocalypse to wipe us out. The chances of one have probably never been higher than now and will, presumably, never be as high again. There is only one thing that we need to fear, it seems, and that is ourselves.

Whether my prediction is true or not, I hope we'll never find out. I am sincerely looking forward to this oddly bright future that we can achieve and I hope that I was able to spread some of that motivation with you. For the time being, keep up the good work and strive towards that glimmer of the future! And always remember: don't be afraid – be curious!

"Live as if you were to die tomorrow. Learn as if you were to live forever."

Mahatma Gandhi

www.ingramcontent.com/pod-product-compliance
Lightning Source LLC
La Vergne TN
LVHW100248210926
848480LV00003B/867
9798232281908